INTRODUCTION TO OPTIMIZATION

Wiley–Interscience Series in Discrete Mathematics and Optimization

Advisory Editors

Ronald L. Graham
AT & T Bell Laboratories, Murray Hill, New Jersey, USA

Jan Karel Lenstra
Centre for Mathematics and Computer Science, Amsterdam, The Netherlands

Robert E. Tarjan
Department of Computer Science, Princeton University, New Jersey, and AT & T Bell Laboratories, Murray Hill, New Jersey, USA

AHLSWEDE and WEGENER—Search Problems
ANDERSON and NASH—Linear Programming in Infinite-Dimensional Spaces
BEALE—Introduction to Optimization
FISHBURN—Interval Orders and Interval Graphs
GONDRAN and MINOUX—Graphs and Algorithms
GOULDEN and JACKSON—Combinatorial Enumeration
GRAHAM, ROTHSCHILD and SPENCER—Ramsey Theory
GROSS and TUCKER—Topological Graph Theory
HALL—Combinatorial Theory
LAWLER, LENSTRA, RINNOOY KAN and SHMOYS—The Traveling Salesman Problem
MINOUX—Mathematical Programming
NEMIROVSKY and YUDIN—Problem Complexity and Method Efficiency in Optimization
PALMER—Graphical Evolution
PLESS—Introduction to the Theory of Error-Correcting Codes
SCHRIJVER—Theory of Linear and Integer Programming
TOMESCU—Problems in Combinatorics and Graph Theory
TUCKER—Applied Combinatorics

INTRODUCTION TO OPTIMIZATION

E. M. L. Beale

Scicon Ltd, Milton Keynes, and Imperial College, London

Edited by

Lynne Mackley

Scicon Ltd, Milton Keynes

A Wiley–Interscience Publication

JOHN WILEY & SONS

Chichester · New York · Brisbane · Toronto · Singapore

Library of Congress Cataloging-in-Publication Data

Beale, E. M. L. (Evelyn Martin Landsdowne)
    Introduction to optimization.

    (Wiley–Interscience series in discrete mathematics
and optimization)
    'A Wiley–Interscience publication.'
    Bibliography: p.
    Includes index.
    1. Mathematical optimization.   I. Mackley, Lynne.
II. Title.   III. Series.
QA402.5.B42   1988   519   87–29613

ISBN 0 471 91760 5
ISBN 0 471 91761 3 (pbk.)

British Library Cataloguing in Publication Data

Beale, E. M. L.
    Introduction to optimization.—(Wiley-
    interscience series in discrete mathematics
    and optimization).
    1. Mathematical optimization
    I. Title   II. Mackley, Lynne
      515   QA402.5

ISBN 0 471 91760 5
ISBN 0 471 91761 3 Pbk

Phototypesetting by Thomson Press (India) Limited, New Delhi
Printed and bound in Great Britain by Anchor Brendon Ltd., Tiptree, Essex

Preface

This book is based upon a series of lecture notes written by Professor E. M. L. Beale FRS for his undergraduate course 'Introduction to Optimization', given at Imperial College where he was a Visiting Professor in Mathematics from 1967 until his death in December 1985. Where necessary, additional material for the chapters on the dual and primal simplex methods has been taken from Professor Beale's book *Mathematical Programming in Practice* (Beale, 1968). In addition, Sections 6.3, 6.4 and 8.1 are based upon the papers 'The current algorithmic scope of mathematical programming systems' (Beale, 1975), and 'The evolution of mathematical programming systems' (Beale, 1985, reproduced by permission of Pergamon Press), whilst the remainder of Chapters 8 and 9 are based on Professor Beale's unpublished paper 'How to apply mathematical programming', which he wrote and continually revised for training courses given by the company Scicon Ltd, of which he was a founder member and where he was for many years Technical Director.

As the title suggests, this book is intended as an introduction to the many topics covered by the heading Optimization, with special emphasis being placed on applications of optimization in industry. Although the lecture notes upon which the book is based were originally written for third-year mathematics undergraduates, no detailed mathematical knowledge is assumed and the book is equally suitable for engineers and computer scientists who are studying options in Operational Research.

The book is divided into three parts. The first concentrates on Unconstrained Optimization and describes some of the main techniques which have been developed to solve problems of this kind. Chapter 2 discusses methods which can be used to optimize functions of only one variable, whilst Chapter 3 considers multi-variable functions. Emphasis is placed on the practical problems of why and how methods succeed or fail, rather than on rigorous proofs about convergence. The second part of the book, Unconstrained Optimization: Linear Programming, describes the methods used to solve linear programming problems and applications of linear programming in industry. The simplex and dual simplex methods are outlined very simply using numerical examples and ways in which the simplex method has been adapted for use on computers are described. Considerable emphasis is placed on the efficient modelling and systematic documentation of linear programming problems. The third part is entitled Unconstrained Optimization: Non-Linear and Discrete, and covers non-linear programming, integer programming and dynamic programming, showing how the techniques of linear programming can be extended to handle non-linearities and discrete entities.

I would like to thank all the people who helped me to edit the book. In particular, I am indebted to Steven Vajda for his advice and encouragement and I would like to thank my colleagues Robert Simons and Bob Hattersley for their advice on technical details about linear programming codes. I am also very grateful to Bev Peberdy for her help in typing the manuscript. Finally, I would like to express my thanks to Scicon Ltd for allowing me time to edit this book.

Lynne Mackley (Scicon Ltd)
August 1987

A biographical memoir of E. M. L. Beale, by M. J. D. Powell, was published in Volume 33 (1987) of *Biographical Memoirs of Fellows of the Royal Society.*

Contents

PART 3   CONSTRAINED OPTIMIZATION: NON-LINEAR AND DISCRETE

1

Introduction

1.1 INTRODUCTION TO OPTIMIZATION

Optimization involves finding the best solution to a problem. Mathematically, this means finding the minimum or maximum of a function of n variables, $f(x_1, \ldots, x_n)$, say, where n may be any integer greater than zero. The function may be unconstrained or it may be subject to certain constraints on the variables of the function, say $g_i(x_1, \ldots, x_n) = b_i$ for $i = 1, \ldots, m$. We will see later that the functions $f(\mathbf{x})$ and $g_i(\mathbf{x})$ usually have some real physical meaning (for example, total cost and profit or capacity and demand restrictions).

Although optimization is used in various branches of applied mathematics and statistics it is particularly associated with *operational research*. Before discussing what operational research involves, we will spend a few minutes distinguishing it from statistics.

We can argue that statistics is concerned with trying to understand what is happening (or what might happen) in an uncertain world full of apparently random phenomena, and that operational research is concerned with deciding what to do about it. If we make this distinction we must add that a practical statistician must then spend some of his time practising operational research, and that a practical operational research worker must spend some of his time practising statistics. We may also add that the quality of the solution to any practical problem may be impaired by dividing the problem up in this way. However, such arbitrary divisions of problems into manageable components are often necessary initial steps towards finding any solutions at all. So I believe that this way of expressing the different approaches of these two major branches of applicable mathematics is of some value.

Operational research is therefore concerned with *decision making*. This can be an instinctive, or at least intuitive, process, but this is not always a satisfactory way to make decisions, particularly those made on behalf of other people, for example by a government department or a commercial organization. We may therefore approach the situation more methodically, and list the alternative possible decisions and their respective advantages and disadvantages. We may then go further and quantify them, and this leads us to make what is generally known as a *mathematical model* of the situation requiring a decision.

Mathematical modelling is at the heart of operational research. This really just means analysing some logical structure that is as simple as possible while still representing the essence of the problem faced by the decision maker.

Let us take a trivial example. Suppose that I take seriously the problem of whether or not to take a raincoat when I leave home in the morning to go to work. Let us suppose that it is not raining, but that I do not want to get wet coming home in the evening. On the other hand, I dislike spending time finding and carrying my raincoat if it is not going to rain. Then I could represent all possible events in the coming day by just two possibilities: wet in the evening or dry in the evening. Then I could try to assess the probabilities of these two alternatives, and I could also try to assess the relative inconveniences of not having a raincoat if it is wet and of carrying one if it is dry. By multiplying the relative inconveniences by the corresponding probabilities, I can decide which action to take.

This is an example of a mathematical model. Like many such models, it does not involve any deep mathematical techniques; but it seems fair to call it mathematical, since it is concerned with expressing the logic of the situation in a formal way.

There are three things that are worth noting about this model, because they are typical of real operational research models. The first is that we do *not* try to make the model as *realistic as possible*. I could easily make the model more realistic. For example, I could consider a range of possible intensities of rainfall, which would affect the amount of inconvenience in not having my raincoat, and I could consider a range of possible temperatures, which would affect the amount of inconvenience in having my raincoat if it is fine. I could also enlarge on the set of possible decisions. For example, if it is raining when I leave work I could run to the bus shelter and plan to stay there until it stops; or I could consider listening to the weather forecast before making my original decision.

Now any or all of these extensions of the model may turn it into a more effective aid to decision making. However, they might equally lead me into a worse state of confusion than my original simple model, perhaps because I have no confidence in any possible way of assessing the probabilities of different temperatures or amounts of rainfall. Just as the problem for which the model is developed is one of finding the best compromise between partially conflicting objectives, so the art of model building itself is one of finding the best compromise between realistically representing the situation and being able to collect data easily and draw conclusions from the results.

The next thing to note is that the use of the model involves *optimization*. I want to know what is best for me, so I choose my decision to minimize the expected inconvenience. In this case the optimization problem is mathematically rather trivial: there are only two possible decisions, so I can compute the expected inconvenience from each, and choose the smaller. However, other problems involve quantitative variables, such as how much of some material to make, or for how long to operate a machine. These problems may require numerical techniques for finding maxima or minima, as well as skill in model building. In this book we will concentrate on these numerical techniques, while giving some thought to model building. It is important for model builders to know a fair amount about optimization techniques, even if they can use existing computer programs to implement them. This is because of the need to compromise in model building between realism and ease of use. We can only do this if we have a fair idea of the numerical problems involved in solving any model.

The third point to be made about this and many other models is that their value is not so much that they give the best answer to the problem—which, of course, is only true to the extent that the model is valid—it is much more that the model provides a convenient *framework* for constructive thought about the problem. In the simplest form of the raincoat model we see two ingredients: the probabilities of certain events and the value of different outcomes in each circumstance. This may lead us on to more elaborate versions of the model if we are dissatisfied with the alternatives offered to us by the simple model.

In practice we often need to see the numerical solution to the model to help us to realize that the data are incomplete or incorrect. This makes the techniques for computing the solutions very important: to solve a real problem we may need to compute the answers to a number of alternative mathematical models, in which case we cannot afford to take too long solving any of them.

1.2   BASIC THEORETICAL NOTIONS

Before we begin studying optimization techniques it is worth spending some time defining the mathematical concepts which are used in the course of this book. In most cases detailed knowledge of these concepts will not be needed and an understanding of the definitions given below will suffice.

The point $\mathbf{x}^*$ in the region R is said to be a *local maximizer* of the function $f(\mathbf{x})$, subject to $\mathbf{x} \in R$, if there exists a small positive number ε such that

$$f(\mathbf{x}^*) \geqslant f(\mathbf{x})$$

for all $\mathbf{x} \in R$ which satisfy $\|\mathbf{x}^* - \mathbf{x}\| < \varepsilon$. The value of $f(\mathbf{x}^*)$ is then the corresponding *local maximum*. The symbol $\in$ is standard mathematical notation, meaning 'is a member of' or 'belongs to'. The norm, or distance measure, is not particularly important. We may use the Euclidean norm, where

$$\|\mathbf{x}\| = \left(\sum_j x_j^2 \right)^{1/2}$$

The point $\mathbf{x}^*$ is a *global maximizer* of the function $f(\mathbf{x})$, where $\mathbf{x} \in R$, if

$$f(\mathbf{x}^*) \geqslant f(\mathbf{x})$$

for *all* $\mathbf{x} \in R$. The value of $f(\mathbf{x}^*)$ is then the *global maximum* of the function $f(\mathbf{x})$ in the region R.

Definitions of *local* and *global minima* follow in the same way, replacing $\geqslant$ by $\leqslant$ in the obvious places. Otherwise we can say that $f(\mathbf{x})$ has a local or global minimum if and only if $-f(\mathbf{x})$ has a local or global maximum.

Certain optimization techniques—generally known as *hill-climbing* techniques—start with an estimate of the global maximum of the function and repeatedly try to improve upon it by finding other points which have a greater function value than the current estimate. The existence of local maxima that are not also global maxima is clearly an undesirable hazard for these optimization techniques, since it is possible that the methods will converge to a local maximizer instead of a global maximizer. Therefore we will consider some important circumstances in which local maxima must also be global maxima.

The most usual such circumstances are connected with the notions of *convexity* and *concavity*. A region R is defined to be a *convex region* if the point

$$(1 - \theta)\mathbf{x}_1 + \theta\mathbf{x}_2 \qquad (0 < \theta < 1)$$

is always in the region, providing the points $\mathbf{x}_1$ and $\mathbf{x}_2$ also belong to it. A *convex function* is one that is never underestimated by linear interpolation, i.e. if

$$\mathbf{x} = (1 - \theta)\mathbf{x}_1 + \theta\mathbf{x}_2 \qquad (0 < \theta < 1)$$

then

$$f(\mathbf{x}) \leqslant (1 - \theta)f(\mathbf{x}_1) + \theta f(\mathbf{x}_2)$$

If this inequality holds with strict inequality, i.e. $<$ rather than $\leqslant$, the function is said to be *strictly convex*. A function $f(\mathbf{x})$ is *concave* if and only if $-f(\mathbf{x})$ is *convex*. Similarly, $f(\mathbf{x})$ is *strictly concave* if and only if $-f(\mathbf{x})$ is *strictly convex*.

Note that a linear function is both convex and concave, and also that a twice-differentiable function $f(x)$ of a single variable is convex if and only if

$$f''(x) \geqslant 0$$

everywhere, where $f''(x)$ is the second derivative of $f(x)$. Similarly, a twice-differentiable function of n variables $f(x_1, \ldots, x_n)$ is convex if its matrix of second partial derivatives is *positive semi-definite* everywhere. In other words,

$$\mathbf{s}^T \mathbf{A} \mathbf{s} \geqslant 0 \qquad \text{(for all vectors } \mathbf{s} \neq 0) \tag{1.2.1}$$

where

$$\mathbf{A} = \begin{pmatrix} \dfrac{\partial^2 f}{\partial x_1^2} & \dfrac{\partial^2 f}{\partial x_1 \partial x_2} & \cdots & \dfrac{\partial^2 f}{\partial x_1 \partial x_n} \\[2mm] \dfrac{\partial^2 f}{\partial x_2 \partial x_1} & \dfrac{\partial^2 f}{\partial x_2^2} & \cdots & \dfrac{\partial^2 f}{\partial x_2 \partial x_n} \\[2mm] \vdots & \vdots & & \vdots \\[2mm] \dfrac{\partial^2 f}{\partial x_n \partial x_1} & \dfrac{\partial^2 f}{\partial x_n \partial x_2} & \cdots & \dfrac{\partial^2 f}{\partial x_n^2} \end{pmatrix}$$

This matrix of second partial derivatives is commonly known as the *Hessian* matrix. A matrix $\mathbf{A}$ is said to be *positive definite* if equation (1.2.1) holds with strict inequality.

The main importance of convexity comes from the following proposition. If the region R is convex and $f(\mathbf{x})$ is a convex function in R, then any local minimizer $\mathbf{x}^*$ of $f(\mathbf{x})$ in R is also a global minimizer.

To prove this, observe that if it were not so, there must exist a point $\mathbf{x}_G$ in R such that

$$f(\mathbf{x}_G) < f(\mathbf{x}^*)$$

So this means that, since R is a convex region, the points given by

$$\mathbf{x} = (1 - \theta)\mathbf{x}^* + \theta \mathbf{x}_G \qquad (0 < \theta < 1)$$

must also belong to R. Also since $f(\mathbf{x})$ is a convex function we know that

$$f(\mathbf{x}) \leqslant (1 - \theta)f(\mathbf{x}^*) + \theta f(\mathbf{x}_G)$$

which in turn implies that $f(\mathbf{x}) < f(\mathbf{x}^*)$ for all θ. However, as θ tends towards zero, so the point $\mathbf{x}$ tends towards $\mathbf{x}^*$, which contradicts the hypothesis that there must exist a positive ε such that $f(\mathbf{x}) \geqslant f(\mathbf{x}^*)$ for all $\mathbf{x}$ such that $\| \mathbf{x} - \mathbf{x}^* \| < \varepsilon$. Hence it must be true that a local minimizer of a convex function in a convex region is also a global minimizer.

An immediate corollary of this is that a local maximizer of a concave function in a convex region must also be a global maximizer. By definition, the actual value of the global maximum must be unique. However, it does not follow that the global maximizer must be unique; multiple global maximizers are possible.

Turning now to more general functions, the *Taylor series* for a continuous function $f(x)$ of a single variable x, with continuous derivatives $f'(x), f''(x)\ldots$ in a given interval $a \leqslant x \leqslant b$, is defined to be

$$f(x) = f(a) + \frac{(x - a)}{1!} f'(a) + \frac{(x - a)^2}{2!} f''(a) + \cdots$$

or, equivalently,

$$f(a + x) = f(a) + \frac{x}{1!} f'(a) + \frac{x^2}{2!} f''(a) + \cdots$$

If $f(\mathbf{x})$ is a function of n variables then the Taylor series, expanded around a point $\mathbf{x}_0$, written in matrix notation becomes

$$f(\mathbf{x}) = f(\mathbf{x}_0) + \mathbf{c}^T(\mathbf{x} - \mathbf{x}_0) + \tfrac{1}{2}(\mathbf{x} - \mathbf{x}_0)^T \mathbf{A}(\mathbf{x} - \mathbf{x}_0) + \cdots$$

where the elements of the vector $\mathbf{c}$ are the first-order partial derivatives $\partial f/\partial x_j$ evaluated at the point $\mathbf{x}_0$, and where the matrix $\mathbf{A}$ is the Hessian matrix, also evaluated at the point $\mathbf{x}_0$.

Note that the Hessian matrix is a *square* matrix because it has the same number of rows and columns. It is also *symmetric*. Writing a_{ij} as the ijth element of $\mathbf{A}$, this means that

$$a_{ij} = a_{ji} \qquad (\text{for all } i, j)$$

or, alternatively,

$$\mathbf{A} = \mathbf{A}^T$$

A *diagonal* matrix is a square matrix which has zero elements everywhere except for the leading diagonal, i.e. $a_{ij} = 0$ for $i \neq j$. A special example of a diagonal matrix is the *unit* matrix, which has all its diagonal elements equal to one. The unit matrix is usually denoted by the symbol **I**. For example,

$$\mathbf{I} = \begin{pmatrix} 1 & 0 & 0 \\ 0 & 1 & 0 \\ 0 & 0 & 1 \end{pmatrix}$$

If the matrix is a square one such that all the elements above the leading diagonal are zero, it is known as a *lower triangular* matrix. Similarly, an *upper triangular* matrix is one where all the elements below the leading diagonal are equal to zero.

The following useful inequality, known as *Cauchy's* inequality, is referred to in Section 4.4 on quasi-Newton methods:

$$\sum_r (x_r^2) \sum_r (y_r^2) \geqslant \left(\sum_r x_r y_r \right)^2$$

or equivalently in matrix notation:

$$\mathbf{x}^T\mathbf{x}\mathbf{y}^T\mathbf{y} \geqslant (\mathbf{x}^T\mathbf{y})^T(\mathbf{x}^T\mathbf{y})$$

The number of ways of choosing r items from a group of n, where order is not important, is called the number of *combinations* of n items taken r at a time and is given by

$$\frac{n!}{(n-r)!r!}$$

The number of possible combinations of n items is 2^n. For example, the possible combinations of the three letters A, B, C are

$$\text{None at all, A, B, C, AB, AC, BC, ABC}$$

The number of *permutations* or arrangements of n items taken r at a time, where order matters, is given by

$$\frac{n!}{(n-r)!}$$

Finally, computer *rounding-off* errors are referred to throughout this book. These errors occur because most real numbers cannot be represented exactly on a

Table 1.2.1

	Fraction	Exact decimal	Binary (to 8 binary places)	Decimal representation of binary
A	2/5	0.4	0.01100110	0.39843750
B	3/5	0.6	0.10011001	0.59765625
A + B	1	1.0	0.11111111	0.99609375

computer. Computers store numbers in base 2 (binary) rather than base 10 (decimal). Each number is represented as a binary fraction, called the mantissa, multiplied by 2 raised to the power of some number, the exponent. The mantissa is a string of binary digits (zero or one) which is truncated after a certain number of places (typically 26). Table 1.2.1 shows an example of the large rounding-off errors that arise when the two fractions 2/5 and 3/5 are stored as binary numbers, truncated after only 8 places.

PART 1

UNCONSTRAINED OPTIMIZATION

2

Introduction to Unconstrained Optimization Techniques

2.1 DIFFERENTIAL CALCULUS

The first part of this book is concerned with *unconstrained optimization*, that is, with minimizing (or maximizing) a function, say $f(\mathbf{x})$, which is not subject to any constraints. The function may be one of continuous variables or discrete variables, or a mixture of the two.

We will start by considering the unconstrained optimization of functions of continuous variables. Now the reader may well imagine that all these problems can be solved using the differential calculus. To minimize a function of n variables $f(x_1, \dots, x_n)$ we simply have to satisfy the n simultaneous equations:

$$\frac{\partial f}{\partial x_j}(x_1, x_2, \dots, x_n) = 0 \qquad (j = 1, \dots, n) \tag{2.1.1}$$

However, the differential calculus does not provide a method for solving such equations if they have no exploitable special structure. On the other hand, if the functions $z_j(\mathbf{x})$ are all real-valued, we can solve the equations

$$z_j(\mathbf{x}) = 0 \qquad (j = 1, \dots, n)$$

by minimizing the sum of the squares of the residuals defined by

$$\sum_{j=1}^{n} (z_j(\mathbf{x}))^2 \tag{2.1.2}$$

We have apparently gone full circle: to minimize the function $f(\mathbf{x})$ we need to solve a set of equations which may be solved by minimizing another function. The question arises whether either of these transformations is of any value.

The calculus approach is useful if the equations can be solved directly—for example, if they are all linear. It is also useful if it enables the dimension of the problem to be reduced. For example, the function may be quadratic in x_1 for fixed values of the other variables. Then we may write

$$f(x_1, x_2, \dots, x_n) = g_0(x_2, \dots, x_n) + g_1(x_2, \dots, x_n)x_1 + g_2(x_2, \dots, x_n)x_1^2$$

This implies that

$$\frac{\partial f}{\partial x_1} = g_1(x_2,\ldots,x_n) + 2g_2(x_2,\ldots,x_n)x_1$$

If $g_2 < 0$ for any values of $x_2,\ldots,x_n$, then we can make f arbitrarily large and negative by making x_1 large enough. Also if $g_2 = 0$ while $g_1 \neq 0$ we can again make f arbitrarily large and negative. Otherwise we find that

$$\min_{x_1} f(\mathbf{x}) = g_0(x_2,\ldots,x_n) - \frac{g_1(x_2,\ldots,x_n)^2}{4g_2(x_2,\ldots,x_n)}$$

We now have an optimization problem in only $n-1$ variables. When this is solved, we can derive the corresponding value of x_1 from

$$x_1 = -\frac{g_1(x_2,\ldots,x_n)}{2g_2(x_2,\ldots,x_n)}$$

2.2 ITERATIVE METHODS

If direct methods fail we may use iterative methods. These are particularly convenient with a computer, because once a single iteration has been performed we need very little extra programming to do an arbitrarily large number of iterations.

Iterative methods are often useful for solving simultaneous equations once they have been transformed into a minimization problem such as equation (2.1.2). Unfortunately, they are not so useful for solving the calculus equations (2.1.1), primarily because stationary points are not necessarily minima and we may iterate towards a saddle point, or even a maximum.

It is better to iterate directly on the function f and use a *valley-descending* method. Valley-descending means finding a minimum of a function $f(\mathbf{x})$ using the following strategy:

(1) Take a trial solution, say $\mathbf{x}_k$;
(2) Find a direction from this trial solution in which $f(\mathbf{x})$ decreases;
(3) Find a point $\mathbf{x}_{k+1}$ in this direction such that $f(\mathbf{x}_{k+1}) < f(\mathbf{x}_k)$;
(4) Repeat the process from this new trial solution.

If the function $f(\mathbf{x})$ is to be maximized, the corresponding strategy where we require $f(\mathbf{x}_{k+1})$ to be greater than $f(\mathbf{x}_k)$ can be called *hill-climbing*. Such a strategy can easily be implemented on a computer. However, when using iterative methods on a computer we must beware of methods that work well on some problems but go into endless loops or just fail on others. The following methods are examples of methods which can fail on certain problems.

Consider the *one variable at a time method*. We begin at some trial solution and alter each variable in turn so as to reduce the function $f(\mathbf{x})$ for fixed values of the other variables. Once all the variables have been changed we can repeat the process. This method works better on some problems than on others, but we may expect that it will always either converge to a local minimum or produce a sequence of trial solutions with $f(\mathbf{x}) \to -\infty$.

It is easy to find a counter-example if $f(\mathbf{x})$ is not differentiable; for example,

$$f(\mathbf{x}) = |x_1 - x_2| + 0.1x_1$$

starting at the origin as the first trial solution. However, Powell (1973) has produced examples of differentiable functions where the method still fails. The first of these examples is

$$\begin{aligned} f(\mathbf{x}) = &- x_1x_2 - x_2x_3 - x_3x_1 + (x_1 - 1)_+^2 \\ &+ (-x_1 - 1)_+^2 + (x_2 - 1)_+^2 + (-x_2 - 1)_+^2 \\ &+ (x_3 - 1)_+^2 + (-x_3 - 1)_+^2 \end{aligned}$$

where

$$(z - c)_+^2 = \begin{cases} 0 & \text{if } z \leqslant c \\ (z - c)^2 & \text{if } z > c \end{cases}$$

It can be verified that if the initial trial solution is the point $(-1 - \varepsilon, 1 + \frac{1}{2}\varepsilon, -1 - \frac{1}{4}\varepsilon)$, the steps of the one variable at a time method (searching along the co-ordinate directions) cycle increasingly closer to six of the edges of the cube with vertices $(\pm 1, \pm 1, \pm 1)$. This failure arises because the search directions increasingly become more nearly orthogonal to the gradient directions at the trial solutions.

Therefore we might consider using the gradient direction as the search direction; in other words, the direction whose jth component is proportional to $\partial f/\partial x_j$. This is called the *method of steepest descent* (or steepest ascent for maximization), because the gradient direction is the direction along which $f(\mathbf{x})$ decreases most rapidly at the point $\mathbf{x}$. Defining the method more formally, if $\mathbf{x}_k$ is the kth trial solution and $\mathbf{g}_k$ is the gradient direction at that point, we choose λ to minimize $f(\mathbf{x}_{k+1})$, where

$$\mathbf{x}_{k+1} = \mathbf{x}_k - \lambda \mathbf{g}_k$$

If the contours of constant value of the function are concentric hyperspheres, then this method produces the optimal solution in one step. With nearly concentric near-hyperspheres it also converges rapidly, but with other functions it may converge very slowly. For example, consider minimizing the function illustrated in Figure 2.2.1:

$$f(\mathbf{x}) = x_1^2 + 1.99x_1x_2 + x_2^2$$

The steepest descent method finds the minimum in one step if we start on the line $x_1 = x_2$. However, Table 2.2.1 shows that if the initial trial solution is a long way from the line $x_1 = x_2$, we progress slowly by zigzagging up the ridge $x_1 = -x_2$.

Such slow convergence may not seem to matter with a powerful computer but, in practice, rounding-off errors tend to affect inefficient methods more seriously than efficient ones (see Section 1.2 for a description of rounding-off errors).

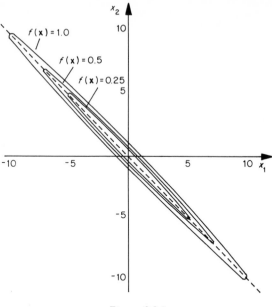

Figure 2.2.1

Table 2.2.1

Iteration number	x_1	x_2	$f(\mathbf{x})$
1	3.00000000	−2.00000000	1.06000000
2	2.49365503	−2.49381167	0.06218708
3	0.17600118	−0.11733412	0.00364833
4	0.14629541	−0.14630460	0.00021404
5	0.01032547	−0.00688365	0.00001256
6	0.00858272	−0.00858326	0.00000074
7	0.00060577	−0.00040384	0.00000004
8	0.00050352	−0.00050355	0.00000000
9	0.00003554	0.00002369	0.00000000

In general, although both of these early methods are intuitively obvious and simple to use, they are best avoided because of their unreliability in certain cases. Therefore with these types of failure in mind we will go on to study some of the more reliable methods for optimizing functions which are now available.

3

One-dimensional Optimization

3.1 FINDING THE ROOTS OF AN EQUATION

Let us now consider the problem of finding a local maximizer of a continuous function $f(x)$ of a scalar argument x. This is important in its own right. It is also required essentially as a subroutine in most numerical methods for optimization in more than one dimension. Finally, it gives some insight into the sort of ideas that go into the development of optimization algorithms.

We start with the slightly simpler problem of finding a root of the equation $f(x) = 0$, where $f(x)$ is again a continuous function of a scalar argument. One popular approach to this problem is *Newton's* method. This involves computing the values of $f(x)$ and $f'(x)$ (the gradient of f at x) for a trial point x_i and then finding an improved estimate x_{i+1} by linear extrapolation (see Figure 3.1.1). Algebraically, this means

$$x_{i+1} = x_i - \frac{f(x_i)}{f'(x_i)}$$

Newton's method must converge if $f(x)$ is either convex or concave and it may converge very rapidly. However, if $f(x)$ is an S-shaped function and we start with a poor trial solution x_0, Newton's method may oscillate with an ever-increasing amplitude. For example, consider the function

$$f(x) = \tan^{-1} x$$

(see Figure 3.1.2) which has the derivative

$$f'(x) = \frac{1}{(1 + x^2)}$$

Table 3.1.1 shows how Newton's method applied to this problem begins to oscillate as the initial trial solution x_0 moves further away from the root at zero. This risk can easily be avoided by using other methods, so Newton's method is usually best avoided for problems in one dimension.

Reliable methods for finding a root involve bracketing the solution by values of x for which $f(x)$ has opposite signs. In practice we nearly always know something about where the relevant root is, so without any loss of generality we may assume

15

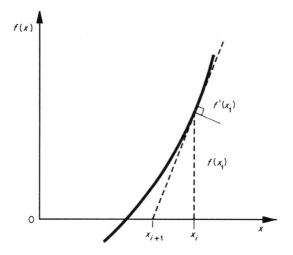

Figure 3.1.1

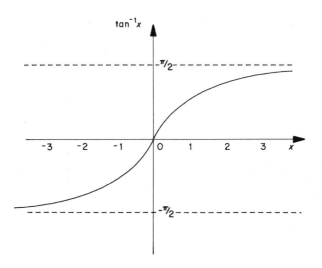

Figure 3.1.2

that we need a root for $x > 0$ and that $f(0) > 0$. We now need to find a positive value of x, say b, such that $f(b) < 0$. If such a value b cannot be found easily and if $f(x)$ is a decreasing function of x, we can use a process of *doubling up*. This may be calculating $f(2^i)$ for $i = 0, 1, 2 \ldots$ until we either find a root or until we find a value of i such that

$$f(2^i) > 0 \quad \text{and} \quad f(2^{i+1}) < 0$$

If we know nothing about $f(x)$ except that it is continuous, there is no general method of finding a value of x such that $f(x) < 0$. This problem is equivalent to

Table 3.1.1

Iteration i	$x_0 = 1.0$ x_i	$f(x_i)$	$x_0 = 1.5$ x_i	$f(x_i)$	$x_0 = 2.0$ x_i	$f(x_i)$
0	1.000	0.785	1.500	0.983	2.000	1.107
1	−0.571	−0.519	−1.694	−1.038	−3.536	−1.295
2	0.117	0.116	2.321	1.164	13.951	1.499
3	−0.001	−0.001	−5.114	−1.378	−279.344	−1.567
4	0.000	0.000	32.296	1.540	122016.999	1.571

finding a global minimum of $f(x)$, which cannot be done without some further restriction. In these circumstances we cannot guarantee that we can show that $f(x) = 0$ has any positive roots, let alone find one.

Therefore let us assume that we have two points, which we denote by a and b, such that

$$f(a)f(b) < 0$$

We now know that there is a root between a and b. The simplest method of narrowing down this range is by *bisection*. We calculate $f(c)$, where $c = \frac{1}{2}(a + b)$ and if

$$f(a)f(c) > 0$$

we replace a by c, whilst if

$$f(a)f(c) < 0$$

we replace b by c. The process is repeated until a point c is found such that $f(a)f(c) = 0$, or until $|a - b|$ is less than some tolerance.

A superior method for well-behaved functions is the *method of false position*, which computes c by linear interpolation (see Figure 3.1.3). In other words,

$$c = a - \frac{f(a)}{f(b) - f(a)}(b - a)$$

Unfortunately, this method is not perfect because if the function $f(x)$ has a second derivative of constant sign between a and b, the new point will always fall on the same side of the root.

We can overcome this by using a *modified method of false position*. Given a and b such that $f(a)f(b) < 0$, we define $f_A = f(a)$ and $f_B = f(b)$ and again calculate c by linear interpolation. In other words,

$$c = a - \frac{f_A}{f_B - f_A}(b - a) \tag{3.1.1}$$

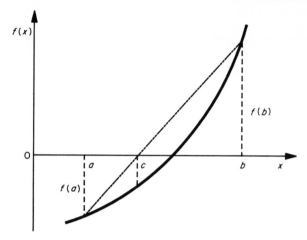

Figure 3.1.3

If $f(a)f(c) = 0$, or if $|a - b|$ is less than some tolerance, then c is the required root. If $f(a)f(c) < 0$, we set b equal to a, with $f_B = f_A$, and a equal to c, with $f_A = f_C$. Otherwise, if $f(a)f(c) > 0$ we replace f_B in equation (3.1.1) by $\frac{1}{2}f_B$ and set a equal to c. The process is repeated.

Table 3.1.2 illustrates how the three methods behave when applied to the function $f(x) = (x + 1)(x + 2)(20x - 11)$, with the initial points $a = 0$ and $b = 1$. The modified method is terminated when $|a - b|$ becomes less than 10^{-6}.

However, even this modified method of false position is not entirely satisfactory. Consider an example where $a = 0$, $f(a) = -1$ and $b = 1$, $f(b) = 10^6 - 1$. Linear interpolation suggests putting $c = 10^{-6}$. If we find that $-1 < f(c) < 0$, the method of false position suggests that we should now

Table 3.1.2

Bisection		Method of false position		Modified method	
c	$f(c)$	c	$f(c)$	c	$f(c)$
0.50000000	−3.75000000	0.28947368	−15.38259950	0.28947368	−15.38259950
0.75000000	19.25000000	0.44700226	−7.29392492	0.54735644	−0.20840038
0.62500000	6.39843750	0.51280852	−2.82759272	0.55423771	0.33646475
0.56250000	1.00097656	0.53704989	−1.00999754	0.54998839	−0.00091744
0.53125000	−1.45349121	0.54554977	−0.35016808	0.54999995	−0.00000402
0.54687500	−0.24623108	0.54847771	−0.12014678	0.55000005	0.00000398
0.55468750	0.37235069	0.54948009	−0.04107644	0.55000000	0.00000000
0.55078125	0.06180787	0.54982253	−0.01402624		
0.54882813	−0.09252414	0.54993943	−0.00478749		
0.54980469	−0.01543633	0.54997933	−0.00163385		
0.55029297	0.02316622	0.54999295	−0.00055757		

interpolate between c and b. The modified method also interpolates between c and b, but with an artificially reduced value of $f(b)$. However, common sense suggests that, when estimating $f(x)$ for values of x only slightly greater than c, the value at a is more relevant than the value at b.

Peters and Wilkinson (1969) recommend a method due to van Wijngaarden *et al.* (1963) which incorporates this idea. This uses both linear interpolation and bisection and works with three points a, b and c. These points are always chosen so that the conditions

$$f_B f_C < 0 \quad \text{and} \quad |f_B| \leqslant |f_C| \tag{3.1.2}$$

hold, where $f_B = f(b)$, $f_C = f(c)$. The value a may be equal to c, or it may be distinct from both b and c.

To initialize the procedure, suppose we have two points u and v such that $f(u)f(v) < 0$. If $|f(u)| \leqslant |f(v)|$, then we set

$$b = u, \quad c = v, \quad a = c$$

whilst if $|f(u)| > |f(v)|$

$$b = v, \quad c = u, \quad a = c$$

A new point d is calculated by linear interpolation between points a and b. In other words,

$$d = b - \frac{f_B}{f_A - f_B}(a - b)$$

However, if the new point lies outside the interval b to $\frac{1}{2}(b + c)$, bisection between b and c is used instead, so that d is set equal to $\frac{1}{2}(b + c)$. The points are then provisionally set to

$$a = b, \quad b = d, \quad c = c$$

with

$$f_A = f_B, \quad f_B = f(d)$$

Conditions (3.1.2) are tested and, if they do not hold, the points a, b and c are redefined as follows. If $f_B f_C > 0$,

$$c = a, \quad f_C = f_A$$

and if $|f_B| > |f_C|$

$$a = b, \quad b = c, \quad c = a$$

with $f_A = f_B$, $f_B = f_C$ and $f_C = f_A$. The process is then repeated. Further refinements of this method are described in Brent (1971).

3.2 MAXIMIZATION

Turning now to the maximization problem, we can bracket a local maximum between two points a and c if

$$a < b < c$$

and

$$f(a) < f(b) > f(c)$$

We can then bisect the larger arm, so that if $|b - a| \geqslant |c - b|$, we put $d = \frac{1}{2}(a + b)$ and calculate $f(d)$. If $f(d) \geqslant f(b)$, we set

$$b = d, \quad c = b$$

Otherwise

$$a = d$$

If $|b - a| < |c - b|$, we set $d = \frac{1}{2}(b + c)$ and proceed similarly. Table 3.2.1 illustrates this method when applied to the function

$$f(x) = \frac{x}{(x^2 + 1)} \tag{3.2.1}$$

with $a = 0.5$, $b = 1.11803398$ and $c = 1.5$.

However, this is a somewhat untidy approach, because the progress made at any step in narrowing down the interval in which the maximum is known to lie depends not only on the relative sizes of the bracket intervals but also on whether the new point d is better than b or not. Furthermore, we may make very good progress at narrowing down the interval at one step and poor progress at the next.

Given three points a, b and c, we can ensure that progress at the next step is the same, whichever potential new bracket is chosen, by choosing d such that $(d - a) = (c - b)$. Furthermore, the actual progress from step to step will remain constant if the initial values a, b and c are chosen so that the ratio $(c - b)/(b - a)$ is a suitable value.

To compute this value, suppose, for convenience, that $(b - a) > (c - b)$ and let $(c - b) = \theta(b - a)$. Then

$$\begin{aligned}
(b - d) &= (b - a) - (d - a) \\
&= (b - a) - (c - b) \\
&= \frac{(1 - \theta)}{\theta}(c - b)
\end{aligned}$$

For constant progress, we want the two potential brackets to have intervals in the same ratio as the original bracket (see Figure 3.2.1). In other words,

$$\frac{(c - b)}{(b - a)} = \frac{(b - d)}{(d - a)} = \frac{(b - d)}{(c - b)} = \theta$$

So we require

$$\frac{(1 - \theta)}{\theta} = \theta$$

or

$$\theta = \frac{\sqrt{(5)} - 1}{2}$$

Table 3.2.1

a	b	c	d	f(b)	f(d)
0.50000000	1.11803398	1.50000000	0.80901699	0.49690400	0.48897724
0.80901699	1.11803398	1.50000000	1.30901699	0.49690400	0.48240453
0.80901699	1.11803398	1.30901699	0.96352549	0.49690400	0.49965505
0.80901699	0.96352549	1.11803398	0.88627124	0.49965505	0.49637793
0.88627124	0.96352549	1.11803398	1.04077973	0.49965505	0.49960086
0.88627124	0.96352549	1.04077973	1.00215261	0.49965505	0.49999884
0.96352549	1.00215261	1.04077973	0.98283905	0.49999884	0.49992510
0.98283905	1.00215261	1.04077973	1.02146617	0.49999884	0.49988725
0.98283905	1.00215261	1.02146617	0.99249583	0.49999884	0.49998582
0.99249583	1.00215261	1.02146617	1.01180939	0.49999884	0.49996554

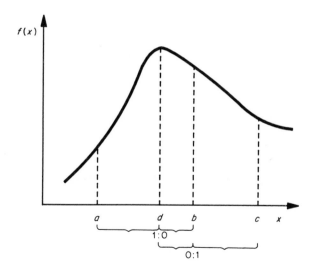

Figure 3.2.1

This method is known as the *golden section search*. Table 3.2.2 demonstrates how it also works when applied to function (3.2.1) with the same initial values of a, b and c.

Table 3.2.3 demonstrates how the interval lengths vary from step to step when the two methods are applied to function (3.2.1). Note in particular that if we calculate the percentage reduction in each interval length, compared with the previous interval length, then these percentages remain the same for the golden section method ('constant progress') but vary considerably from step to step for the bisection method.

Just as there are methods for solving $f(x) = 0$ that are appreciably faster than bisection for linear functions, there ought to be methods that are appreciably

Table 3.2.2

a	b	c	d	$f(b)$	$f(d)$
0.50000000	1.11803398	1.50000000	0.88196602	0.49690400	0.49608181
0.88196602	1.11803398	1.50000000	1.26393204	0.49690400	0.48659105
0.88196602	1.11803398	1.26393204	1.02786407	0.49690400	0.49981123
0.88196602	1.02786407	1.11803398	0.97213593	0.49981123	0.49980041
0.97213593	1.02786407	1.11803398	1.06230584	0.49981123	0.49908808
0.97213593	1.02786407	1.06230584	1.00657770	0.49981123	0.49998925
0.97213593	1.00657770	1.02786407	0.99342230	0.49998925	0.49998911
0.99342230	1.00657770	1.02786407	1.01470867	0.49998925	0.49994670
0.99342230	1.00657770	1.01470867	1.00155326	0.49998925	0.49999940
0.99342230	1.00155326	1.00657770	0.99844674	0.49999940	0.49999940

Table 3.2.3

BISECTION		GOLDEN SECTION	
$|c - a|$	Change in length (%)	$|c - a|$	Change in length (%)
0.69098301	30.90	0.61803398	38.20
0.50000000	27.64	0.38196602	38.20
0.30901699	38.20	0.23606796	38.20
0.23176274	25.00	0.14589805	38.20
0.15450850	33.33	0.09016991	38.20
0.07725425	50.00	0.05572814	38.20
0.05794069	25.00	0.03444177	38.20
0.03862712	33.33	0.02128637	38.20
0.02897034	25.00	0.01315540	38.20
0.01931356	33.33	0.00813096	38.19

faster than golden section for approximately quadratic functions. However, I regret that I am not yet convinced by any that I have seen. If the one-dimensional optimization is used as a subroutine within a multi-dimensional optimization problem, an exact answer is not needed, and it is probably satisfactory to fit a quadratic function through the data at the points a, b and c. If the maximum of this fitted quadratic is greater than $f(b) + \varepsilon$ (where ε is a small positive number), the point d can be taken as the location of this maximum. If the fitted quadratic has a maximum value less than or equal to $f(b) + \varepsilon$, the process is stopped. This method can obviously fail to locate the maximum (for example, it will always stop if $(c - b) = (b - a)$ and $f(a) = f(c)$), but this does not matter in this context.

To find a guaranteed good approximation to a local maximum with appreciably less work than the golden section method seems difficult. We must avoid taking the new point d very near to either a or c, otherwise we may make a negligible reduction in the interval. We must also avoid taking d very near to b, since rounding-off errors may then cause a misappreciation of the sign of $f(b) - f(d)$.

Sometimes it is convenient to compute first derivatives as well as function values. If we have two points a and b such that $a < b$, $f'(a) > 0$ and either $f(b) < f(a)$ or $f'(b) < 0$, then a local maximum lies between a and b. We could look for this maximum by solving the equation $f'(x) = 0$, but this ignores valuable information about the function value. Furthermore, it does not guarantee finding a function value as good as $f(a)$ if we find that $f'(a) > 0$, $f'(b) > 0$ but $f(b) < f(a)$.

Therefore when derivative information is easily available it is preferable to fit a cubic approximation to $f(x)$ using the function and derivative values. The following method, based on a method in Fletcher and Powell (1963), uses this approach but there is a subtle point of some numerical interest concerning the solution of quadratic equations.

Suppose we are given the values f_0 and f_h of a differentiable function $f(x)$ for $x = x_0$ and $x = x_0 + h$, respectively ($h > 0$), with the corresponding first derivatives f'_0 and f'_h. If $f'_0 > 0$ and either $f_h < f_0$ or $f'_h < 0$, then the function has a local maximum at some point $x = x_0 + \theta h$, where $0 < \theta < 1$. We approximate θ by maximizing a cubic approximation to $f(x)$ fitted to known data.

If the cubic approximation to $f(x_0 + \theta h)$ is denoted by $a_0 + a_1\theta + a_2\theta^2 + a_3\theta^3$, then

$$f_0 = a_0$$
$$f_h = a_0 + a_1 + a_2 + a_3$$
$$hf'_0 = a_1$$
$$hf'_h = a_1 + 2a_2 + 3a_3$$

Solving these equations for the unknown values of a_2 and a_3, we obtain

$$a_2 = -h(t_1 + f'_0)$$

and

$$a_3 = h(\tfrac{2}{3}t_1 + \tfrac{1}{3}f'_0 + \tfrac{1}{3}f'_h)$$

where

$$t_1 = f'_0 - \frac{3}{h}(f_h - f_0) + f'_h$$

The values of θ for which $f(x_0 + \theta h)$ are stationary are the roots of the quadratic equation

$$a_1 + 2a_2\theta + 3a_3\theta^2 = 0$$

so

$$\theta = \frac{-a_2 \pm \sqrt{(a_2^2 - 3a_1a_3)}}{3a_3}$$

To establish the appropriate sign, note that if $a_3 < 0$ we want the larger root, while if $a_3 > 0$ we want the smaller one. Therefore in both cases the negative sign is appropriate, and the formula is given by

$$\theta = -\left[\frac{a_2 + \sqrt{(a_2^2 - 3a_1a_3)}}{3a_3}\right] \tag{3.2.2}$$

If $a_2 < 0$ this formula can produce serious round-off errors, so in this case we can derive an alternative formula by multiplying top and bottom by

$$a_2 - \sqrt{(a_2^2 - 3a_1a_3)}$$

so that

$$\theta = \frac{-a_1}{[a_2 - \sqrt{(a_2^2 - 3a_1a_3)}]} \tag{3.2.3}$$

This is a good formula if $a_2 < 0$. We now note that

$$a_2^2 - 3a_1a_3 = h^2(t_1^2 - f_0'f_h')$$

Therefore if we now substitute for a_1, a_2 and a_3 in equations (3.2.2) and (3.2.3) we obtain

$$\theta = \frac{f_0'}{[t_2 + \sqrt{(t_1^2 - f_0'f_h')}]} \qquad \text{if } t_2 > 0$$

and

$$\theta = \frac{t_2 - \sqrt{(t_1^2 - f_0'f_h')}}{t_1 + t_2 + f_h'} \qquad \text{if } t_2 < 0$$

where $t_2 = t_1 + f_0'$.

A single application of this method will often produce an adequate solution; for example, a single application to the example (3.2.1) considered earlier gives the approximation 1.00317663. However, it is not entirely obvious what to do if the solution is not adequate. It may be best to switch to bisections of the interval to make guaranteed progress, or to the method by van Wijngaarden et al. if the point with the smallest value of $f'(x)$ also has the best function value of all the points investigated so far.

4

Multi-dimensional Optimization

4.1 INTRODUCTION

There has been an immense amount of work in the past on numerical methods for unconstrained optimization of functions of n variables where $n > 1$. Most of this has been concerned with finding local optima of functions that are twice-differentiable, so we will concentrate on this.

You might reasonably hope that the work in this area has led to at least one method which is effective for all problems of this type, but unfortunately that will never be possible, since the choice of method must depend on the answers to questions such as:

(1) How easy is it to calculate function values?
(2) Can we easily calculate the gradient vector at any solution; in other words, can we easily calculate all first derivatives?
(3) Can we easily calculate the Hessian matrix at any trial solution; in other words, can we easily calculate all second derivatives? ·
(4) Can we afford to store an $n \times n$ matrix representing the true or estimated Hessian or, alternatively, its inverse?
(5) Is the Hessian sparse, that is, are many of its elements zero? If so, should we take advantage of this fact?
(6) Does the problem have any special features that we should exploit to make the solution easier to find?

Because there are no easy answers to such questions industry and commerce will always require specialized help to solve practical problems of this nature. The choice of method to solve each problem is likely to be better if the strengths and weaknesses of the more successful existing methods are understood.

So we will study three existing methods which use quite different techniques and have different advantages and limitations. The first, *Newton's* method, requires the calculation of second derivatives and can be modified to incorporate line searches to improve its reliability. The other two methods calculate first and not second derivatives, but nevertheless converge in a finite number of steps when the function is quadratic. The first of these, known as the *conjugate gradient* method, has the advantage for large problems that it does not require the storage of any large $n \times n$ matrices. The other method, known as a *quasi-Newton* method, builds up an explicit approximation to the Hessian matrix and is generally more effective than conjugate gradients when the computer has sufficient storage.

4.2 EXTENSIONS OF NEWTON'S METHOD

The simplest useful general approximation to a twice-differentiable function in the neighbourhood of a minimum or maximum is a quadratic. Therefore methods have been developed to converge in a finite number of steps when the function to be optimized, known as the *objective function*, is quadratic. They have also been adapted to converge reliably when the objective function is not quadratic.

In some ways the most natural approach is therefore to take a trial solution, say $\mathbf{x}_0$, and to calculate the function value and its first and second derivatives at this point. We can then use these values to approximate the function $f(\mathbf{x})$ by a second-order Taylor series:

$$f(\mathbf{x}) \simeq f(\mathbf{x}_0) + \mathbf{c}^T(\mathbf{x} - \mathbf{x}_0) + \tfrac{1}{2}(\mathbf{x} - \mathbf{x}_0)^T \mathbf{A}(\mathbf{x} - \mathbf{x}_0)$$

where $\mathbf{c}$ and $\mathbf{A}$ are the gradient vector and the Hessian matrix, respectively, evaluated at the point $\mathbf{x}_0$.

The gradient vector of the approximating quadratic function, in other words, the vector whose components are $\partial f / \partial x_j$, can be written as $\mathbf{g}(\mathbf{x})$, where

$$\mathbf{g}(\mathbf{x}) = \mathbf{c} + \mathbf{A}(\mathbf{x} - \mathbf{x}_0) \qquad (4.2.1)$$

To prove this, note that

$$\frac{\partial f}{\partial x_i} = c_i + \tfrac{1}{2}\left[2a_{ii}(x_i - x_{i0}) + \sum_{j \neq i} a_{ji}(x_j - x_{j0}) + \sum_{j \neq i} a_{ij}(x_j - x_{j0}) \right]$$

$$= c_i + \sum_j a_{ij}(x_j - x_{j0})$$

The point where the gradient vector vanishes is then approximated by solving the equations

$$\mathbf{c} + \mathbf{A}(\mathbf{x} - \mathbf{x}_0) = 0$$

so that

$$\mathbf{x} = \mathbf{x}_0 - \mathbf{A}^{-1}\mathbf{c}$$

This leads us to *Newton's method* for solving the equations $\partial f / \partial x_j = 0$, for all j. For a trial solution $\mathbf{x}_{i-1}$, evaluate the gradient vector at that point, $\mathbf{g}_{i-1}$, and the Hessian matrix $\mathbf{A}_{i-1}$ and set

$$\mathbf{x}_i = \mathbf{x}_{i-1} - \mathbf{A}_{i-1}^{-1}\mathbf{g}_{i-1}$$

There are various things wrong with this method, but the more serious of these can be corrected quite easily. The first problem is that since the step is based entirely on the local behaviour of the function $f(\mathbf{x})$ in the neighbourhood of $\mathbf{x}_{i-1}$

there is no guarantee that $f(\mathbf{x}_i) < f(\mathbf{x}_{i-1})$. In other words, there is no strong expectation that the method will converge. We can, however, modify the method to incorporate a line search, finding a scalar $\alpha_i > 0$ to minimize $f(\mathbf{x}_i)$, where $\mathbf{x}_i$ is given by

$$\mathbf{x}_i = \mathbf{x}_{i-1} - \alpha_i \mathbf{A}_{i-1}^{-1} \mathbf{g}_{i-1}$$

We probably do not want to spend much effort on optimizing α_i but just want a better trial solution than $f(\mathbf{x}_{i-1})$. We will naturally start by trying the value $\alpha_i = 1$.

However, this method still will not work if $\mathbf{A}_{i-1}$ is singular (so that its inverse does not exist) or if the stationary value of the quadratic approximation to $f(\mathbf{x})$ is not a minimum. In other words, the method will not work unless $\mathbf{A}_{i-1}$ is *positive definite*. This difficulty can also be overcome. The most elegant way was first proposed by Levenberg (1944) and applied to Newton's method by Goldfeld *et al.* (1966). A constant λ is added to all diagonal elements of $\mathbf{A}$ so that we choose α_i to minimize $f(\mathbf{x}_i)$, where $\mathbf{x}_i$ is given by

$$\mathbf{x}_i = \mathbf{x}_{i-1} - \alpha_i (\mathbf{A}_{i-1} + \lambda \mathbf{I})^{-1} \mathbf{g}_{i-1}$$

and where λ is chosen to make $(\mathbf{A}_{i-1} + \lambda \mathbf{I})$ positive definite.

Note that if we put $\alpha_i = 1$ we are simply finding the point that minimizes our quadratic approximation to $f(\mathbf{x})$ subject to the constraint that we only want to move a certain distance from the current trial solution $\mathbf{x}_{i-1}$. This statement can be justified in an elementary way as follows.

Suppose that we are concerned with a quadratic function $Q(\mathbf{x})$

$$Q(\mathbf{x}) = Q(\mathbf{x}_0) + \mathbf{c}^T(\mathbf{x} - \mathbf{x}_0) + \tfrac{1}{2}(\mathbf{x} - \mathbf{x}_0)^T \mathbf{A}(\mathbf{x} - \mathbf{x}_0)$$

and that $F(\mathbf{x})$ is the function formed by replacing the Hessian matrix $\mathbf{A}$ by $(\mathbf{A} + \lambda \mathbf{I})$. In other words,

$$F(\mathbf{x}) = Q(\mathbf{x}) + \tfrac{1}{2}\sum_j \lambda(x_j - x_{j0})^2$$

Then if $\mathbf{x}^*$ is the point that minimizes $F(\mathbf{x})$ we know that

$$F(\mathbf{x}^*) \leqslant F(\mathbf{x})$$

for all $\mathbf{x}$, so that

$$Q(\mathbf{x}^*) \leqslant Q(\mathbf{x}) + \tfrac{1}{2}\sum_j \lambda(x_j - x_{j0})^2 - \tfrac{1}{2}\sum_j \lambda(x_j^* - x_{j0})^2$$

In other words, $Q(\mathbf{x}^*) \leqslant Q(\mathbf{x})$ for all $\mathbf{x}$ such that

$$\sum_j \lambda(x_j - x_{j0})^2 \leqslant \sum_j \lambda(x_j^* - x_{j0})^2$$

This last condition means that $\mathbf{x}^*$ is not only the point that minimizes the adjusted function $F(\mathbf{x})$ but it is also the local minimizer of the original function

$Q(\mathbf{x})$ for all points $\mathbf{x}$ that are no further than $\mathbf{x}^*$ from $\mathbf{x}_0$ in some metric or measure of distance.

In fact, the method often works better if we implicitly rescale the variables so that the positive diagonal elements of $\mathbf{A}_{i-1}$ are all equal. This means that instead of adding a constant to these values we multiply them by a number slightly greater than one. Any extra additions needed to make the matrix positive definite can be determined during the process of matrix inversion.

I had better explain this last remark, even though it involves a digression. To invert the positive definite symmetric matrix $\mathbf{A}$, we form the matrix equation

$$\mathbf{v} = \mathbf{Au}$$

which can be written in algebraic notation as

$$v_i = \sum_{j=1}^{n} a_{ij}u_j \qquad (i = 1, \ldots, n)$$

and then solve these equations for the u_j in terms of the v_i. We start by using the first equation to solve for u_1 in terms of v_1 and the remaining u_j. It turns out that we can preserve symmetry by writing the v_i with a negative sign when they reach the right-hand side of the equation. So after one step we have

$$u_1 = \bar{a}_{11}(-v_1) + \bar{a}_{12}u_2 + \cdots + \bar{a}_{1n}u_n$$
$$v_2 = \bar{a}_{21}(-v_1) + \bar{a}_{22}u_2 + \cdots + \bar{a}_{2n}u_n$$
$$v_n = \bar{a}_{n1}(-v_1) + \bar{a}_{n2}u_2 + \cdots + \bar{a}_{nn}u_n$$

where

$$\bar{a}_{11} = -1/a_{11}$$
$$\bar{a}_{1j} = \bar{a}_{j1} = -a_{1j}/a_{11} \qquad (j \neq 1)$$
$$\bar{a}_{ij} = a_{ij} - \frac{a_{i1}a_{1j}}{a_{11}} \qquad (i, j \neq 1) \tag{4.2.2}$$

At the next step we substitute for u_2 in terms of $v_2, v_1, u_3, \ldots, u_n$ in the same way, and after n steps we will have equations of the form

$$u_i = \sum_{j=1}^{n} \bar{a}_{ij}(-v_j) \qquad (i = 1, \ldots, n)$$

The coefficients $\bar{a}_{ij}$ are then just the negatives of the coefficients of $\mathbf{A}^{-1}$.

Now $\mathbf{A}$ is positive definite if and only if $\mathbf{x}^T\mathbf{A}\mathbf{x} > 0$ for all $\mathbf{x} \neq 0$, and we note that we can rewrite this as

$$\mathbf{x}^T\mathbf{A}\mathbf{x} = \sum_{i=1}^{n} \sum_{j=1}^{n} a_{ij}x_i x_j = \frac{1}{a_{11}}\left(\sum_{j=1}^{n} a_{1j}x_j\right)^2 + \sum_{i=2}^{n} \sum_{j=2}^{n} \bar{a}_{ij}x_i x_j$$

where $\bar{a}_{ij}$ is given by equation (4.2.2). So the expression is positive only if $a_{11} > 0$ and the $(n-1) \times (n-1)$ matrix with coefficients $\bar{a}_{ij}$ ($i_{ij} \geqslant 2$) is positive definite.

Thus the whole matrix is positive definite if and only if the diagonal elements on which we operate, i.e. the $\bar{a}_{pp}$ at the time we solve for u_p in terms of $v_1, \ldots, v_p$,

$u_{p+1}, \ldots, u_n$, are all positive. So if a negative diagonal element is found, we modify the matrix by adding a large enough quantity to this diagonal element to make it small and positive. This turns out to be equivalent to adding the same quantity to the corresponding element of the original matrix $\mathbf{A}$.

Research workers in optimization methods have recently taken more interest in these second-order methods, but I still do not like them on both practical and theoretical grounds. The practical grounds are that the writing of a computer program to calculate the second derivatives correctly can be a painful and expensive process. The theoretical grounds are that the method makes essentially no use of all the information about the function that has been gained in previous iterations.

4.3   CONJUGATE GRADIENTS

The method of conjugate gradients was introduced by Hestenes and Stiefel (1952), but first applied to general optimization problems by Fletcher and Reeves (1964). We will begin by developing the theory of the method and then show that it leads to a practical computational procedure.

Let us assume that we are minimizing a quadratic objective function

$$f(x_1, \ldots, x_n) = f(\mathbf{x}_0) + \mathbf{c}^T(\mathbf{x} - \mathbf{x}_0) + \tfrac{1}{2}(\mathbf{x} - \mathbf{x}_0)^T \mathbf{A}(\mathbf{x} - \mathbf{x}_0) \qquad (4.3.1)$$

by a sequence of line searches in directions $\boldsymbol{\xi}_i$. That is, we choose α_i to minimize $f(\mathbf{x}_i)$, where $\mathbf{x}_i$ is given by

$$\mathbf{x}_i = \mathbf{x}_{i-1} + \alpha_i \boldsymbol{\xi}_i \qquad (4.3.2)$$

A set of search directions $\boldsymbol{\xi}_i$ are said to be *mutually conjugate with respect to A* if

$$\boldsymbol{\xi}_i^T \mathbf{A} \boldsymbol{\xi}_j = 0 \qquad (i \neq j) \qquad (4.3.3)$$

We now show that if a set of mutually conjugate non-null vectors are taken as successive search directions along which the quadratic function $f(\mathbf{x})$ is minimized, then

(1) After any number of steps we have the minimum value of the function in the hyperplane through our current trial solution spanned by the search directions used; and

(2) After, at most, n steps, we have the unconstrained minimum, where n is the number of variables the function depends on.

These results are intuitively obvious when $\mathbf{A}$ is a unit matrix and, by equation (4.3.3), the conjugate directions become orthogonal directions. To prove the results generally we note that after r steps, by equation (4.3.2), we can write

$$\mathbf{x}_r = \mathbf{x}_0 + \sum_{i=1}^{r} \alpha_i \boldsymbol{\xi}_i \qquad (4.3.4)$$

so that substituting $\mathbf{x}_r$ for $\mathbf{x}$ in equation (4.3.1) we obtain

$$f(\mathbf{x}_r) = f(\mathbf{x}_0) + \sum_{i=1}^{r} \alpha_i \mathbf{c}^T \boldsymbol{\xi}_i + \tfrac{1}{2} \sum_{i=1}^{r} \sum_{j=1}^{r} \alpha_i \alpha_j \boldsymbol{\xi}_i^T \mathbf{A} \boldsymbol{\xi}_j$$

However, since the search directions $\boldsymbol{\xi}_i$ are mutually conjugate this reduces to

$$f(\mathbf{x}_r) = f(\mathbf{x}_0) + \sum_{i=1}^{r} (\alpha_i \mathbf{c}^T \boldsymbol{\xi}_i + \tfrac{1}{2}\alpha_i^2 \boldsymbol{\xi}_i^T \mathbf{A} \boldsymbol{\xi}_i) \tag{4.3.5}$$

Hence we can optimize with respect to each α_i separately without the appropriate value of one being dependant on the trial values of the others. So after, at most, n steps we must have found the unconstrained minimum of $f(x_1,\ldots,x_n)$. This suggests that we should try to use a sequence of search directions which are mutually conjugate. Therefore we now consider how we might derive such a set of mutually conjugate search directions given an arbitrary set of n linearly independent directions $\boldsymbol{\eta}_i$, $i = 1,\ldots,n$.

Suppose, for the moment, that we actually know the Hessian matrix $\mathbf{A}$. We begin by setting $\boldsymbol{\xi}_1 = \boldsymbol{\eta}_1$ and then we let

$$\boldsymbol{\xi}_i = \boldsymbol{\eta}_i + \sum_{j=1}^{i-1} b_{ij} \boldsymbol{\xi}_j \qquad (i = 2, 3, \ldots, n) \tag{4.3.6}$$

where the coefficients b_{ij} are chosen to make the search direction $\boldsymbol{\xi}_i$ mutually conjugate to all the previous search directions. In other words,

$$\boldsymbol{\xi}_i^T \mathbf{A} \boldsymbol{\xi}_k = 0 \qquad (k = 1, \ldots, i-1)$$

Using equation (4.3.6), this means that the coefficients b_{ij} must satisfy the equations

$$\boldsymbol{\eta}_i^T \mathbf{A} \boldsymbol{\xi}_k + \sum_{j=1}^{i-1} b_{ij} \boldsymbol{\xi}_j^T \mathbf{A} \boldsymbol{\xi}_k = 0 \qquad (k = 1, \ldots, i-1)$$

In principle, for each i, this defines $i-1$ simultaneous linear equations for the $i-1$ unknowns b_{ij}. However, if we use the fact that the search directions $\boldsymbol{\xi}_1, \ldots, \boldsymbol{\xi}_{i-1}$ are themselves mutually conjugate we see that these reduce to $i-1$ simple equations:

$$b_{ij} = \frac{-\boldsymbol{\eta}_i^T \mathbf{A} \boldsymbol{\xi}_j}{\boldsymbol{\xi}_j^T \mathbf{A} \boldsymbol{\xi}_j} \qquad (j = 1, \ldots, i-1) \tag{4.3.7}$$

We now come to the crucial step that turns all this into something of practical relevance. We can find out what we need to know about the Hessian matrix $\mathbf{A}$ by observing the differences in the gradient vector $\mathbf{g}(\mathbf{x})$ at different points along our search directions. Specifically, if we consider two points $\mathbf{x}_i'$ and $\mathbf{x}_i' + \mu_i \boldsymbol{\xi}_i$, where $\mu_i > 0$, and if we define the gradient difference vector $\mathbf{q}_i$ by the equation

$$\mathbf{q}_i = \mathbf{g}(\mathbf{x}_i' + \mu_i \boldsymbol{\xi}_i) - \mathbf{g}(\mathbf{x}_i') \tag{4.3.8}$$

we see from equation (4.2.1) that

$$\mathbf{q}_i = (\mathbf{c} + \mathbf{A}(\mathbf{x}_i' + \mu_i \xi_i - \mathbf{x}_0)) - (\mathbf{c} + \mathbf{A}(\mathbf{x}_i' - \mathbf{x}_0))$$
$$= \mu_i \mathbf{A} \xi_i \tag{4.3.9}$$

This means that the condition of mutual conjugacy, $\xi_i^T \mathbf{A} \xi_j = 0$, reduces to the condition

$$\xi_i^T \mathbf{q}_j = 0 \qquad (i \neq j)$$

In other words, the new search direction ξ_i should be orthogonal to all previous gradient difference vectors $\mathbf{q}_j$. Equation (4.3.7) can now be rewritten as

$$b_{ij} = -\frac{\eta_i^T \mathbf{q}_j}{\xi_j^T \mathbf{q}_j} \qquad (j = 1, \ldots, i-1)$$

It is conventional to apply this formula with $\mathbf{x}_i' = \mathbf{x}_{i-1}$ and $\mu_i = \alpha_i$, so that we study the changes in the gradient vectors at the optimum points reached on successive line searches. However, this can cause serious rounding-off errors in the calculation of $\mathbf{q}_i$ if $\mathbf{x}_i$ proves to be very near $\mathbf{x}_{i-1}$, because we are then calculating the difference between two nearly equal vectors. A larger value of μ_i is then appropriate. It is therefore possible to generate mutually conjugate search directions without computing the Hessian matrix $\mathbf{A}$ explicitly. Also, by the theorem at the start of Section 4.3, we know that after, at most, n searches in these directions we will obtain the unconstrained minimum, providing $f(\mathbf{x})$ is a quadratic function and the line searches are all exact.

In fact further simplifications to the formulae occur if we define the n linearly independent directions η_i to be $-\mathbf{g}_{i-1}$ in all cases. First, note that the gradient vector $\mathbf{g}_r$ at any point $\mathbf{x}_r$ is orthogonal to all previous search directions:

$$\mathbf{g}_r^T \xi_j = 0 \qquad (j = 1, \ldots, r-1) \tag{4.3.10}$$

To show this we use equation (4.3.5) and set $\partial f / \partial \alpha_j = 0$, giving

$$\mathbf{c}^T \xi_j + \alpha_j \xi_j^T \mathbf{A} \xi_j = 0 \qquad (j = 1, \ldots, r) \tag{4.3.11}$$

However, from equations (4.2.1) and (4.3.4) we can write

$$\mathbf{g}_r = \mathbf{c} + \sum_{i=1}^{r} \alpha_i \mathbf{A} \xi_i$$

so that using the fact that the search directions are mutually conjugate and the Hessian matrix $\mathbf{A}$ is symmetric, we obtain

$$\mathbf{g}_r^T \xi_j = \mathbf{c}^T \xi_j + \alpha_j \xi_j^T \mathbf{A} \xi_j \qquad (j = 1, \ldots, r-1)$$

which is equal to zero by equation (4.3.11), as required.

Using equation (4.3.6), with $\eta_i = -\mathbf{g}_{i-1}$, equation (4.3.10) in turn implies that

$$\mathbf{g}_i^T \mathbf{q}_j = 0 \qquad (j = 1, \ldots, i-1)$$

so that

$$b_{ij} = 0 \qquad (j = 1, \ldots, i - 2)$$

and

$$b_{i,i-1} = \frac{\mathbf{g}_{i-1}^T \mathbf{q}_{i-1}}{\boldsymbol{\xi}_{i-1}^T \mathbf{q}_{i-1}}$$

Therefore when the directions $\boldsymbol{\eta}_i$ are taken to be $-\mathbf{g}_{i-1}$, the formula for the next search direction reduces to the very simple form

$$\boldsymbol{\xi}_i = -\mathbf{g}_{i-1} + \frac{\mathbf{g}_{i-1}^T \mathbf{q}_{i-1}}{\boldsymbol{\xi}_{i-1}^T \mathbf{q}_{i-1}} \boldsymbol{\xi}_{i-1}$$

Because each new search direction only depends upon data from the previous step, this means that the amount of computer storage needed to operate the method increases linearly and not quadratically as the number of variables n increases.

This derivation shows the crucial importance of taking the first search direction in the gradient direction, $\boldsymbol{\xi}_1 = -\mathbf{g}_0$. If we do not do this, there is no reason to believe that the vector $\mathbf{q}_1$ is parallel to the hyperplane in which we have already optimized $f(\mathbf{x})$. Therefore there is no reason to believe that b_{i1} is zero. It follows that if we use this method on a non-quadratic objective function we should not continue to use the standard formula indefinitely. Sooner or later we should restart with a search in the gradient direction. The most logical way to do this, recommended by Powell (1977), is to test for the approximate orthogonality of the vectors $\mathbf{g}_i$ and $\mathbf{g}_{i-1}$. This requires the storage of one more vector, but it prevents futile attempts to build up approximately conjugate directions that have been fatally undermined by non-quadratic behaviour at an early stage.

One other property of the method is worth noting. If the objective function is quadratic, the coefficients $b_{i,i-1}$ are always non-negative when the linearly independent $\boldsymbol{\eta}_i$ are taken to be the gradient vectors $-\mathbf{g}_{i-1}$. Hence if the actual formula applied to a non-quadratic function gives a negative value for $b_{i,i-1}$ it seems best to start again with a gradient direction.

4.4   QUASI-NEWTON METHODS

If we can afford the computer storage space, *quasi-Newton methods* are generally an improvement on the performance of the conjugate gradient method. We begin with an arbitrary positive definite approximation to the Hessian matrix $\mathbf{A}$ (or alternatively its inverse $\mathbf{A}^{-1}$) and gradually improve upon it using the information provided by the gradient difference vectors (4.3.8).

This approach was pioneered by Davidon (1959) and an improved version, known as the *DFP method*, is given in Fletcher and Powell (1963). In 1970 Broyden, Fletcher, Goldfarb and Shanno independently proposed a minor variant of the DFP method which is now known as the *BFGS update formula*. It seems to be rather more satisfactory than the original method and it is this version which we will now study.

Before formally defining the BFGS method we will again try to justify it with some relevant theory. First, however, we need to define some more notation. Let

(a_{ij}) denote the elements of $\mathbf{A}$, the true Hessian matrix,

$(a_{ij}^{(k)})$ denote the elements of $\mathbf{A}_k$, the kth approximation to the Hessian matrix,

$(q_j^{(k)})$ denote the elements of $\mathbf{q}_k$, the kth gradient difference vector,

and finally let

$$Q_k = \mathbf{x}^T \mathbf{A}_k \mathbf{x} = \sum_i \sum_j a_{ij}^{(k)} x_i x_j$$

Now suppose that we want to minimize a quadratic function of n variables:

$$f(\mathbf{x}) = f(\mathbf{x}_0) + \mathbf{c}^T(\mathbf{x} - \mathbf{x}_0) + \tfrac{1}{2}(\mathbf{x} - \mathbf{x}_0)^T \mathbf{A}(\mathbf{x} - \mathbf{x}_0)$$

and that we have chosen an initial search direction ξ_1. Without any real loss of generality we can imagine that we have rotated axes so that this is in the x_1 direction. In other words,

$$\xi_1 = (1, 0, 0, \ldots, 0)^T$$

If we now compute the vector $\mathbf{q}_1$ as in the method of conjugate gradients we know from equation (4.3.9) that

$$\mathbf{q}_1 = \mu_1 \mathbf{A} \xi_1$$

so that

$$a_{1j} = a_{j1} = q_j^{(1)}/\mu_1 \tag{4.4.1}$$

using the notation defined earlier. So if we start with an arbitrary positive definite approximation $\mathbf{A}_0$ to the true Hessian matrix $\mathbf{A}$, we might consider using equation (4.4.1) to improve it. In other words, we might try forming $\mathbf{A}_1$ where

$$a_{ij}^{(1)} = \begin{cases} q_j^{(1)}/\mu_1 & \text{for } i, j = 1 \\ a_{ij}^{(0)} & \text{for } i, j > 1 \end{cases}$$

However, this might be a fairly devastating thing to try, as we have no guarantee that the matrix $\mathbf{A}_1$ formed in this way will be positive definite. So we try forming

$$Q_0 = \sum_i \sum_j a_{ij}^{(0)} x_i x_j$$

which we can expand as

$$Q_0 = a_{11}^{(0)} x_1^2 + 2 \sum_{j=2}^{n} a_{1j}^{(0)} x_1 x_j + \sum_{i=2}^{n} \sum_{j=2}^{n} a_{ij}^{(0)} x_i x_j$$

By completing the square for x_1 this can be rewritten as

$$Q_0 = \frac{1}{a_{11}^{(0)}} \left(\sum_j a_{1j}^{(0)} x_j \right)^2 + \sum_{i=2}^{n} \sum_{j=2}^{n} \bar{a}_{ij}^{(0)} x_i x_j$$

where

$$\bar{a}_{ij}^{(0)} = a_{ij}^{(0)} - a_{i1}^{(0)}a_{1j}^{(0)}/a_{11}^{(0)}$$

We can now try forming $\mathbf{A}_1$ by using equation (4.4.1) to update the elements $a_{1j}^{(0)}$ in the first term of the expression for Q_0. Therefore

$$Q_1 = \mathbf{x}^T\mathbf{A}_1\mathbf{x} = \frac{1}{\mu_1 q_1^{(1)}}\left(\sum_j q_j^{(1)}x_j\right)^2 + \sum_{i=2}^n \sum_{j=2}^n \bar{a}_{ij}^{(0)}x_ix_j$$

This formula can be written in matrix notation as

$$Q_1 = \mathbf{x}^T\mathbf{A}_1\mathbf{x} = \mathbf{x}^T\mathbf{A}_0\mathbf{x} - \frac{\mathbf{x}^T(\mathbf{A}_0\mathbf{e}_1)(\mathbf{A}_0\mathbf{e}_1)^T\mathbf{x}}{\mathbf{e}_1^T\mathbf{A}_0\mathbf{e}_1} + \frac{\mathbf{x}^T\mathbf{q}_1\mathbf{q}_1^T\mathbf{x}}{\mu_1\mathbf{e}_1^T\mathbf{q}_1} \tag{4.4.2}$$

with $\mathbf{e}_1 = (1,0,0,\ldots,0)^T$.

Is the new updated matrix $\mathbf{A}_1$ positive definite as required? A necessary and sufficient condition for any matrix $\mathbf{H}$ to be positive definite is that it can be expressed as

$$\mathbf{H} = \mathbf{L}\mathbf{L}^T$$

where, given $\mathbf{H}, \mathbf{L}$ is a unique, lower triangular matrix. So if we write $\mathbf{A}_0 = \mathbf{L}_0\mathbf{L}_0^T$, $\mathbf{a} = \mathbf{L}_0^T\mathbf{x}$ and $\mathbf{b} = \mathbf{L}_0^T\mathbf{e}_1$, equation (4.4.2) becomes

$$Q_1 = \frac{\mathbf{a}^T\mathbf{a}\mathbf{b}^T\mathbf{b} - (\mathbf{a}^T\mathbf{b})(\mathbf{a}^T\mathbf{b})^T}{\mathbf{b}^T\mathbf{b}} + \frac{(\mathbf{x}^T\mathbf{q}_1)(\mathbf{x}^T\mathbf{q}_1)^T}{\mu_1\mathbf{e}_1^T\mathbf{q}_1}$$

which means that

$$Q_1 > \frac{(\mathbf{x}^T\mathbf{q}_1)(\mathbf{x}^T\mathbf{q}_1)^T}{\mu_1\mathbf{e}_1^T\mathbf{q}_1} > 0, \qquad \text{providing } q_1^{(1)} > 0$$

using Cauchy's inequality (see Section 1.2). So $\mathbf{A}_1$ is positive definite, as required, whilst $q_1^{(1)} > 0$. This must happen if $f(\mathbf{x})$ is quadratic, unless $f(\mathbf{x})$ decreases indefinitely with an arbitrarily large step in the x_1 direction. However, $q_1^{(1)}$ may be non-positive when we are optimizing a more general function, in which case $\mathbf{A}_0$ cannot usefully be updated.

Therefore when $q_1^{(1)} > 0$, $\mathbf{A}_1$ is positive definite and we can write it as $\mathbf{L}_1\mathbf{L}_1^T$, where the first column of $\mathbf{L}_1$ must consist of the elements

$$\frac{q_j^{(1)}}{(\mu_1 q_1^{(1)})^{1/2}} \qquad q_1^{(1)} > 0$$

if the $a_{1j}^{(1)}$ are to have the correct values. Since there are no requirements on the other elements of $\mathbf{L}_1$, it is natural to make them the same as the corresponding elements of $\mathbf{L}_0$.

When the search direction is the more general direction ξ_k, it is easy to show that the general update formula is given by

$$\mathbf{A}_k = \mathbf{A}_{k-1} - \frac{(\mathbf{A}_{k-1}\boldsymbol{\xi}_k)(\mathbf{A}_{k-1}\boldsymbol{\xi}_k)^T}{\boldsymbol{\xi}_k^T \mathbf{A}_{k-1}\boldsymbol{\xi}_k} + \frac{\mathbf{q}_k \mathbf{q}_k^T}{\mu_k \boldsymbol{\xi}_k^T \mathbf{q}_k} \tag{4.4.3}$$

which again should not be used in the exceptional situation when $\boldsymbol{\xi}_k^T \mathbf{q}_k \leqslant 0$.

We now study the properties of this generalized formula. First, it is easy to show, using Cauchy's inequality as we did earlier, that providing our initial approximation $\mathbf{A}_0$ is positive definite then $\mathbf{A}_k$ will also be positive definite. Second, since the true Hessian matrix $\mathbf{A}$ satisfies

$$\mathbf{q}_k = \mu_k \mathbf{A}\boldsymbol{\xi}_k \qquad k \geqslant 1$$

we would like our updated approximations $\mathbf{A}_k$ to satisfy this relationship too. Post-multiplying by $\boldsymbol{\xi}_k$ we deduce that

$$\mathbf{A}_k\boldsymbol{\xi}_k = \mathbf{A}_{k-1}\boldsymbol{\xi}_k - \frac{\mathbf{A}_{k-1}\boldsymbol{\xi}_k \boldsymbol{\xi}_k^T \mathbf{A}_{k-1}^T \boldsymbol{\xi}_k}{\boldsymbol{\xi}_k^T \mathbf{A}_{k-1}\boldsymbol{\xi}_k} + \frac{\mathbf{q}_k \mathbf{q}_k^T \boldsymbol{\xi}_k}{\mu_k \boldsymbol{\xi}_k^T \mathbf{q}_k}$$

$$= \mathbf{q}_k/\mu_k \quad \text{as required}$$

In other words,

$$\mathbf{A}_k\boldsymbol{\xi}_k = \mathbf{A}\boldsymbol{\xi}_k$$

Now suppose we take a new search direction $\boldsymbol{\xi}_{k+1}$ and apply the same procedure. It is clear that then we will have the relationship

$$\mathbf{A}_{k+1}\boldsymbol{\xi}_{k+1} = \mathbf{A}\boldsymbol{\xi}_{k+1}$$

However, will we have distorted the information incorporated after the previous line search? In other words, will $\mathbf{A}_{k+1}\boldsymbol{\xi}_k = \mathbf{A}\boldsymbol{\xi}_k$? To answer this question in general terms we post-multiply our updating formula by an arbitrary search direction $\boldsymbol{\xi}_j$ where $j < k$. We then see that

$$\mathbf{A}_k\boldsymbol{\xi}_j = \mathbf{A}_{k-1}\boldsymbol{\xi}_j - \frac{\mathbf{A}_{k-1}\boldsymbol{\xi}_k \boldsymbol{\xi}_k^T \mathbf{A}_{k-1}^T \boldsymbol{\xi}_j}{\boldsymbol{\xi}_k^T \mathbf{A}_{k-1}\boldsymbol{\xi}_k} + \frac{\mathbf{q}_k \mathbf{q}_k^T \boldsymbol{\xi}_j}{\mu_k \boldsymbol{\xi}_k^T \mathbf{q}_k}$$

$$= \mathbf{A}_{k-1}\boldsymbol{\xi}_j \quad \text{if} \quad \boldsymbol{\xi}_k^T \mathbf{A}_{k-1}\boldsymbol{\xi}_j = 0 \quad \text{and} \quad \mathbf{q}_k^T \boldsymbol{\xi}_j = 0$$

However,

$$\mathbf{q}_k^T \boldsymbol{\xi}_j = \mu_k \boldsymbol{\xi}_k^T \mathbf{A}\boldsymbol{\xi}_j$$

Hence we see that all past information is preserved correctly provided that each search direction $\boldsymbol{\xi}_k$ is chosen to be conjugate to all previous search directions with respect to $\mathbf{A}_{k-1}$. We now show that we can achieve this by setting

$$\boldsymbol{\xi}_k = -\mathbf{A}_{k-1}^{-1}\mathbf{g}_{k-1}$$

This is a very natural search direction, since it would point straight towards the optimum solution if $\mathbf{A}_{k-1}$ were the true Hessian $\mathbf{A}$. The proof that this works depends on the fact that

$$\mathbf{g}_{k-1}^T \boldsymbol{\xi}_j = 0 \qquad (j < k)$$

which implies that $\boldsymbol{\xi}_k^T \mathbf{A}_{k-1} \boldsymbol{\xi}_j = 0$, as required.

This is the basis of quasi-Newton methods for unconstrained optimization. Their main advantage over conjugate gradient methods is that they avoid the need for periodic restarts at which much information is lost. Their main disadvantage is that they require the storage of an $n \times n$ matrix. However, the apparent further disadvantage that they require a matrix inversion at each iteration is illusory, since if we start with an expression for $\mathbf{A}_0^{-1}$ we can transform the updating formula into an updating formula for $\mathbf{A}_k^{-1}$. Writing $\mathbf{A}_k^{-1}$ as $\mathbf{H}_k$ it can be shown that

$$\mathbf{H}_k = \mathbf{H}_{k-1} + \left(1 + \frac{\mathbf{q}_k^T \mathbf{H}_{k-1} \mathbf{q}_k}{\mu_k \mathbf{q}_k^T \boldsymbol{\xi}_k}\right)\left(\frac{\mu_k \boldsymbol{\xi}_k \boldsymbol{\xi}_k^T}{\mathbf{q}_k^T \boldsymbol{\xi}_k}\right)$$
$$- \left(\frac{\boldsymbol{\xi}_k \mathbf{q}_k^T \mathbf{H}_{k-1} + \mathbf{H}_{k-1} \mathbf{q}_k \boldsymbol{\xi}_k^T}{\mathbf{q}_k^T \boldsymbol{\xi}_k}\right)$$

PART 2

CONSTRAINED OPTIMIZATION: LINEAR PROGRAMMING

5

The Simplex Method for Linear Programming

5.1 INTRODUCTION

In Part 1 we considered the optimization of unconstrained functions. We will now go on to study techniques for optimizing linear functions which are subject to linear inequality constraints. In other words, we will be looking at problems of the form

$$\min \sum_{j=1}^{n} c_j x_j$$

subject to the constraints

$$\sum_{j=1}^{n} a_{ij} x_j \leqslant b_i \qquad (i = 1, \ldots, m) \tag{5.1.1}$$

where c_j and a_{ij} are constants or known data and the x_j are variables (assumed to be non-negative). Note that we have not restricted ourselves to only consider less-than-or-equal-to constraints, since any greater-than-or-equal-to constraint $g_k(\mathbf{x}) \geqslant b_k$ can be rewritten as

$$-g_k(\mathbf{x}) \leqslant -b_k$$

whilst an equality constraint $g_k(\mathbf{x}) = b_k$ can be represented by two inequality constraints

$$g_k(\mathbf{x}) \geqslant b_k \quad \text{and} \quad g_k(\mathbf{x}) \leqslant b_k$$

The study of problems of the form (5.1.1) is known as *linear programming* (which is perhaps a rather unfortunate name since most people associate 'programming' with writing computer programs). In Chapter 9 we will see some examples of linear programming problems, and in particular we will be looking at some typical problems which are encountered in industry. However, for now we will see how we can actually solve linear programming problems, and we will start with the particular method of solution known as the *simplex* method.

This method was devised by Dantzig in 1947, although the first published reference was in 1951. It is easiest to introduce the simplex method through a small-scale numerical example of a particular type of linear programming (LP) problem, known as a *single-mix blending* problem.

5.2 THE SINGLE-MIX BLENDING PROBLEM

Suppose that we want to produce an animal food mix which will satisfy certain dietary restrictions on the amounts of vitamins, minerals or carbohydrates contained in the mixture. We can make this food mix from various foodstuffs (or *raw materials*) such as sago flour, meat meal or enriched skimmed milk. Our problem is to decide which of these raw materials we want to include in the mixture, and in what proportions, so that we produce the cheapest mix possible which satisfies the dietary constraints. This problem, known as a single-mix blending problem because we are only trying to produce one mixture, can be formulated mathematically as follows:

Subscripts

i A quality of the mixture, e.g. a property such as protein, fat or vitamin A. There may be a restriction or *quality constraint* on the amount of this property allowed in the mixture. There are I qualities allowed, i.e. $i = 1, \ldots, I$.

j A raw material which can be used in the mixture, e.g. sago flour, meat meal or enriched skimmed milk. There are J different raw materials, i.e. $j = 1, \ldots, J$.

Sets

S_L The set of qualities with lower limits, i.e. quality i_1 belongs to this set if the mixture must contain at least a certain amount of quality i_1.

S_U The set of qualities with upper limits, i.e. quality i_2 belongs to this set if the mixture must not contain more than a certain amount of quality i_2.

Constants

A_{ij} The amount of quality i in each unit of raw material j (tonnes/tonne).

C_j The cost of raw material j (£/tonne).

Q_{Li} The lower limit on the amount of quality i allowed in the mixture (tonnes/tonne).

Q_{Ui} The upper limit on the amount of quality i allowed in the mixture (tonnes/tonne).

No other data are assumed relevant and, in particular, no other costs are considered important.

Decision variables

x_j The proportion of raw material j in the mixture.

The problem is to minimize the total cost of producing a tonne of the mixture, subject to the restrictions on some of the qualities, i.e.

Minimize
Objective function

$$\sum_{j=1}^{J} C_j x_j$$ The total cost of producing a tonne of the mixture

subject to
Total tonnage constraint

$$\sum_{j=1}^{J} x_j = 1$$

Quality constraints

$$\sum_{j=1}^{J} A_{ij} x_j \geqslant Q_{Li}   \quad (i \in S_L)$$

$$\sum_{j=1}^{J} A_{ij} x_j \leqslant Q_{Ui}   \quad (i \in S_U)$$

Non-negativity constraints

$$x_j \geqslant 0   \quad (j = 1, \ldots, J)$$

As a numerical example consider the following problem. The raw materials are sago flour, meat meal and enriched skimmed milk and the qualities of the mixture are protein, fat, carbohydrate and fibre and ash. So $I = 4$ and $J = 3$. The remaining data for the example are summarized in Table 5.2.1.

Table 5.2.1

Raw material		Sago flour	Meat meal	Enriched skimmed milk
Quality	Quality restriction (%)	Amount of quality in raw material (%)		
Protein	$\geqslant 20$	0	50	30
Fat	$\geqslant 10$	0	10	20
Carbohydrate	—	100	0	40
Fibre and ash	$\leqslant 20$	0	40	10
Cost of raw material (£/tonne)		40	50	70

Thus the set S_L consists of the qualities protein and fat, whilst the set S_U consists of the quality fibre and ash. Working in tens of £/tonne and in tenths to keep the numbers as small integers, the problem becomes:

Choose non-negative x_1, x_2 and x_3 to minimize

$$C = 4x_1 + 5x_2 + 7x_3$$

subject to

$$
\begin{aligned}
x_1 + x_2 + x_3 &= 1 \\
5x_2 + 3x_3 &\geqslant 2 \\
x_2 + 2x_3 &\geqslant 1 \\
4x_2 + x_3 &\leqslant 2
\end{aligned}
\tag{5.2.1}
$$

To solve this problem, we start by multiplying all greater-than-or-equal-to constraints by -1 to turn them into less-than-or-equal-to constraints. We then reduce it to one containing only equality constraints in non-negative variables by introducing *slack variables*, representing the difference between the left- and right-hand sides of each inequality. So we have

$$
\begin{aligned}
C = 4x_1 + 5x_2 + 7x_3 \\
x_1 + x_2 + x_3 &= 1 \\
- 5x_2 - 3x_3 + s_2 &= -2 \\
- x_2 - 2x_3 + s_3 &= -1 \\
4x_2 + x_3 + s_4 &= 2
\end{aligned}
$$

We now solve the equations for some variables in terms of others, and write the slacks in terms of the decision variables without any arithmetic:

$$
\begin{aligned}
C &= 4x_1 + 5x_2 + 7x_3 \\
0 &= 1 - x_1 - x_2 - x_3 \\
s_2 &= -2 + 5x_2 + 3x_3 \\
s_3 &= -1 + x_2 + 2x_3 \\
s_4 &= 2 - 4x_2 - x_3
\end{aligned}
$$

If we now solve the equation with zero on the left-hand side for x_3 and substitute for it in the other equations we have

$$
\begin{aligned}
C &= 7 - 3x_1 - 2x_2 \\
x_3 &= 1 - x_1 - x_2 \\
s_2 &= 1 - 3x_1 + 2x_2 \\
s_3 &= 1 - 2x_1 - x_2 \\
s_4 &= 1 + x_1 - 3x_2
\end{aligned}
\tag{5.2.2}
$$

Note that this 'solved' form of the equations is fully equivalent to the original problem (5.2.1)—we have just rewritten the equations. The variables on the left-hand sides of equations (5.2.2) are called *basic* variables and those on the right-hand sides are called *non-basic* variables. We now consider a *trial solution* to the problem in which all the non-basic variables are set to zero, i.e. $x_3 = 1$, $s_2 = 1$, $s_3 = 1$ and $s_4 = 1$.

This trial solution is said to be *feasible* since all the decision and slack variables are non-negative, as required. However, it is not the optimal solution to the problem since the expression for C contains negative elements. In other words, we can further decrease the value of the objective function C by increasing the value of x_1 or x_2 away from zero.

We therefore apply a valley-descending technique to reduce C and increase x_1, keeping x_2 equal to zero. The values of the basic variables are then given by

$$x_3 = 1 - x_1$$
$$s_2 = 1 - 3x_1$$
$$s_3 = 1 - 2x_1$$
$$s_4 = 1 + x_1 \qquad (5.2.3)$$

Since $\partial C / \partial x_1$ is a negative constant, we obviously want to make x_1 as large as possible to reduce the value of C. However, we can see from equations (5.2.3) that the trial solution will only remain feasible if $x_1 \leqslant \frac{1}{3}$, since s_2 is negative when $x_1 > \frac{1}{3}$.

If we now put $x_1 = \frac{1}{3}$ we must consider whether x_2 can usefully be increased. The answer is clearly yes for this example, but in general this is much less obvious when a basic variable is up against its bound in the trial solution. We could find that in order to increase x_2 we have to decrease x_1 to ensure that the solution remains feasible, and this may or may not further reduce the value of C. Alternatively, we could find that the direct effect of increasing x_2 increases the value of C but is still worth while because it allows us to further increase the value of x_1.

These difficulties can be avoided by changing the sets of basic and non-basic variables so that we do not have basic variables up against their individual bounds. In other words, we make a basic variable non-basic when it is driven to zero, replacing it with the non-basic variable which has just been increased. So we change the sets of basic and non-basic variables at each step of the iterative solution process.

Therefore writing x_1 in terms of s_2 using the third equation of (5.2.2) we substitute this expression for x_1 in the objective function and other constraints. The equations now read

$$C = 6 + s_2 - 4x_2$$
$$x_3 = \tfrac{2}{3} + \tfrac{1}{3}s_2 - \tfrac{5}{3}x_2$$
$$x_1 = \tfrac{1}{3} - \tfrac{1}{3}s_2 + \tfrac{2}{3}x_2$$
$$s_3 = \tfrac{1}{3} + \tfrac{2}{3}s_2 - \tfrac{7}{3}x_2$$
$$s_4 = \tfrac{4}{3} - \tfrac{1}{3}s_2 - \tfrac{7}{3}x_2 \qquad (5.2.4)$$

Note that C has decreased from 7 in equation (5.2.2) to 6 in equation (5.2.4). However, it is still not optimal because x_2 has a negative coefficient in the expression for C. So we increase x_2 away from zero until s_3 becomes negative. Therefore, making x_2 basic and s_3 non-basic, we obtain

$$C = \tfrac{38}{7} - \tfrac{1}{7}s_2 + \tfrac{12}{7}s_3$$
$$x_3 = \tfrac{3}{7} - \tfrac{1}{7}s_2 + \tfrac{5}{7}s_3$$
$$x_1 = \tfrac{3}{7} - \tfrac{1}{7}s_2 - \tfrac{2}{7}s_3$$
$$x_2 = \tfrac{1}{7} + \tfrac{2}{7}s_2 - \tfrac{3}{7}s_3$$
$$s_4 = 1 - s_2 + s_3$$

The expression for C still contains a negative coefficient, so we now increase s_2 away from zero until s_4 becomes zero. Making s_2 basic and s_4 non-basic the equations become

$$C = \tfrac{37}{7} + \tfrac{1}{7}s_4 + \tfrac{11}{7}s_3$$
$$x_3 = \tfrac{2}{7} + \tfrac{1}{7}s_4 + \tfrac{4}{7}s_3$$
$$x_1 = \tfrac{2}{7} + \tfrac{1}{7}s_4 - \tfrac{3}{7}s_3$$
$$x_2 = \tfrac{3}{7} - \tfrac{2}{7}s_4 - \tfrac{1}{7}s_3$$
$$s_2 = 1 - s_4 + s_3 \qquad\qquad (5.2.5)$$

So the minimum cost is £52.86 per tonne. Because all our decision variables x_1, x_2 and x_3 ended up as basic variables the optimal mixture contains all of our raw materials in the proportions $\tfrac{2}{7}$ sago flour, $\tfrac{3}{7}$ meat meal and $\tfrac{2}{7}$ enriched skimmed milk.

This is a *global* optimum, and follows mathematically from the fact that we are minimizing a convex function in a convex region. However, it can be proved from first principles. Since the coefficients of the non-basic variables in the objective function are non-negative, C is obviously minimized subject to the constraints that the non-basic variables must be non-negative, and the additional constraints on the basic variables are also satisfied.

Note the coefficients in the final expression for the objective function. These are known as *reduced costs*, and indicate the minimum cost of increasing the value of a decision variable and, in the case of a slack, the cost of tightening a constraint or, conversely, the benefit from relaxing it. For example, the reduced cost of the slack variable s_4 is $\tfrac{1}{7}$. This means that if we decrease the fibre and ash requirement by $10\,\theta\%$ (where θ is a small amount), while satisfying the other constraints, the cost of the mixture will increase by £$(10/7)\,\theta$ per tonne. Unfortunately the calculations we have performed so far do not tell us for how large a range of values of θ these costs hold. However, we will return to this point in *parametric programming* in Section 7.4.

Problems with two non-basic variables can be represented geometrically. For example, Figure 5.2.1 illustrates the previous problem. The figure relates to equations (5.2.2) when x_1 and x_2 were the non-basic variables. The lines correspond to the points where the variables are equal to zero and the *feasible region* is the pentagon ABCDE in the bottom left-hand corner of the figure. The contours of constant value of the objective function

$$C = 7 - 3x_1 - 2x_2$$

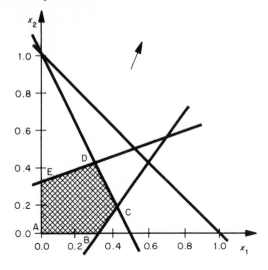

Figure 5.2.1

are parallel straight lines, so they can be summarized by a line drawn normally to all of them in the direction of decreasing values of the objective function (see the arrow in the top right-hand corner of the figure). In principle, we can solve this problem by rotating the figure so that the arrow is pointing vertically upward and then find the highest point in the feasible region. On this basis the optimal solution is obviously given by the point D. Thus if the problem only has two non-basic coefficients, we can solve the problem by this geometrical method. Unfortunately the accuracy of our solution will depend on the scale of the figure, and this method will not provide us with the reduced costs information which is often as useful as the solution itself.

From equations (5.2.2)–(5.2.5) we can see that the simplex method starts at some vertex of the polyhedron and at each stage moves to a neighbouring one with a better value of the objective function. The method terminates when no such vertex can be found. It is therefore a *finite* iterative method because there are only a finite number of vertices.

There is a theoretical loophole in this argument: there may be more variables with a trial value of zero than can be made non-basic. This is known as *degeneracy*, and can be avoided by making an arbitrarily small perturbation in the constraints. However, the practical problems arising from degeneracy are essentially problems of numerical precision, and rather different techniques are used to overcome them.

5.3   THE ALGEBRA OF THE SIMPLEX METHOD

The numerical solution process illustrated in the previous section can be expressed algebraically. We need symbols for the variables which indicate if they are currently basic or non-basic, and also symbols for the numerical coefficients in the current set of equations. Rather than use subscripted subscripts, such as x_{j_i}, we use capital X's to represent a permutation of the original variables x. So we can write

$$x_0 = \bar{a}_{00} - \sum_{j=1}^{n-m} \bar{a}_{0j} X_{m+j}$$

$$X_i = \bar{a}_{i0} - \sum_{j=1}^{n-m} \bar{a}_{ij} X_{m+j} \qquad (i = 1, \ldots, m) \qquad (5.3.1)$$

This set of equations is known as a *tableau*. $X_1, \ldots, X_m$ represent the current basic variables and $X_{m+1}, \ldots, X_n$ the non-basic variables. Note that the objective function C has been written as x_0, since it behaves like any basic variable except that it is not necessarily restricted to non-negative values. More specifically, it is conventional to maximize x_0, so C has been replaced by $-x_0$. The symbol $\bar{a}_{ij}$ is used for the coefficients to indicate that they are a permutation of the original coefficients a_{ij}. The negative signs before these coefficients are traditional, and they arise naturally if we think of the original formulation in matrix notation as

$$\mathbf{Ax} = \mathbf{b}$$

with x_0 included in the vector $\mathbf{x}$. If we now partition $\mathbf{A}$ and $\mathbf{x}$ we obtain

$$(\mathbf{B}|\mathbf{N})(\mathbf{x_B}|\mathbf{x_N})^T = \mathbf{b}$$

where $\mathbf{x_B}$ and $\mathbf{x_N}$ are the vectors formed by the basic and non-basic variables, respectively, and where $\mathbf{B}$ and $\mathbf{N}$ are the columns of $\mathbf{A}$ associated with these variables. Pre-multiplying by $\mathbf{B}^{-1}$ we obtain

$$\mathbf{x_B} = \mathbf{B}^{-1}\mathbf{b} - \mathbf{B}^{-1}\mathbf{N}\mathbf{x_N}$$

So the tableau elements $\bar{a}_{ij}$ are therefore the elements of $\mathbf{B}^{-1}\mathbf{N}$.

The problem is to maximize x_0 subject to the constraints that the basic and non-basic variables $X_1, \ldots, X_n$ must all be non-negative. At the current trial solution $X_i = \bar{a}_{i0}$, for $i = 1, \ldots, m$, and $X_i = 0$, for $i = m + 1, \ldots, n$. We assume that $\bar{a}_{i0} \geqslant 0$, for $i = 1, \ldots, m$, or else we do not have a feasible trial solution. Methods for finding a first feasible solution are discussed in Chapter 6.

If all the coefficients $\bar{a}_{0j}$ (the reduced costs) are positive the trial solution is optimum. Otherwise we select a negative one, say $\bar{a}_{0q}$. Theoretically it does not matter which negative reduced cost we take, but in practice we usually take the most negative. The corresponding variable X_{m+q} is then chosen to enter the basis. To determine the variable to leave the basis we compute the ratios

$$\frac{\bar{a}_{i0}}{\bar{a}_{iq}} \qquad (5.3.2)$$

for all i such that $a_{iq} > 0$. If $\bar{a}_{p0}/\bar{a}_{pq}$ is the smallest ratio, X_p is the variable chosen to leave the basis, since this ratio determines the maximum possible increase in X_{m+q} that keeps the trial solution feasible.

When defining any algorithm we must consider whether anomalous situations can arise, and if so what we should do about them. If there is no i such that $\bar{a}_{iq}$ in equation (5.3.2) is greater than zero, the solution is *unbounded*. In that case the objective function can be made arbitrarily large by making X_{m+q} arbitrarily large and varying the basic variables accordingly. If there is more than one i with the same smallest ratio then we have the phenomenon of degeneracy, noted at the end of Section 5.2. In practice, it is usual to choose the eligible row which maximizes $\bar{a}_{pq}$. More specifically, we follow Harris (1973) in choosing the row giving the largest value of $\bar{a}_{pq}$ subject to the constraint that no basic variable can become less than $-T_{\mathrm{OLB}}$, where T_{OLB} is a small tolerance. This tends to improve the numerical precision of the calculations.

Having determined the variables to enter and leave the basis, the formulae for transforming the tableau can be derived by writing

$$x_0 = \bar{a}_{00} - \sum_{\substack{j=1 \\ j \neq q}}^{n-m} \bar{a}_{0j} X_{m+j} - \bar{a}_{0q} X_{m+q}$$

$$X_p = \bar{a}_{p0} - \sum_{\substack{j=1 \\ j \neq q}}^{n-m} \bar{a}_{pj} X_{m+j} - \bar{a}_{pq} X_{m+q}$$

$$X_i = \bar{a}_{i0} - \sum_{\substack{j=1 \\ j \neq q}}^{n-m} \bar{a}_{ij} X_{m+j} - \bar{a}_{iq} X_{m+q} \qquad (i \neq p)$$

so that rewriting the second equation in terms of X_{m+q} we obtain

$$X_{m+q} = \frac{\bar{a}_{p0}}{\bar{a}_{pq}} - \sum_{\substack{j=1 \\ j \neq q}}^{n-m} \frac{\bar{a}_{pj}}{\bar{a}_{pq}} X_{m+j} - \frac{1}{\bar{a}_{pq}} X_p$$

If we now substitute for X_{m+q} in the other equations we can write the new tableau as

$$x_0 = \bar{a}'_{00} - \sum_{j=1}^{n-m} \bar{a}'_{0j} X'_{m+j}$$

$$X'_i = \bar{a}'_{i0} - \sum_{j=1}^{n-m} \bar{a}'_{ij} X'_{m+j} \qquad (i = 1, \ldots, m)$$

where $X'_p = X_{m+q}$, $X'_{m+q} = X_p$ and where $X'_i = X_i$ for $i \neq p$ and $i \neq m+q$. The new coefficients are given by

$$\bar{a}'_{pq} = 1/\bar{a}_{pq}$$
$$\bar{a}'_{pj} = \bar{a}_{pj}/\bar{a}_{pq} \qquad (j \neq q)$$
$$\bar{a}'_{iq} = -\bar{a}_{iq}/\bar{a}_{pq} \qquad (i \neq p)$$
$$\bar{a}'_{ij} = \bar{a}_{ij} - \bar{a}_{pj}\bar{a}_{iq}/\bar{a}_{pq} \qquad (i \neq p, j \neq q) \qquad (5.3.3)$$

These formulae for the new tableau entries can be written in the more convenient form

$$\bar{a}'_{pq} = 1/\bar{a}_{pq}$$
$$\bar{a}'_{pj} = \bar{a}_{pj}\bar{a}'_{pq} \qquad (j \neq q)$$
$$\bar{a}'_{iq} = -\bar{a}_{iq}\bar{a}'_{pq} \qquad (i \neq p)$$
$$\bar{a}'_{ij} = \bar{a}_{ij} - \bar{a}_{iq}\bar{a}'_{pj} \qquad (i \neq p, j \neq q) \tag{5.3.4}$$

The second form, given by equations (5.3.4), is usually used in numerical calculations because it involves fewer division and multiplication operations. However, the first form (5.3.3) does emphasize the key role played by the element $\bar{a}_{pq}$, which is known as the *pivot*. The qth column is called the *pivotal column* and the pth row the *pivotal row*.

6

Further Details of the Simplex Method

6.1 BASIC METHOD OF FINDING A FIRST FEASIBLE SOLUTION

In Chapter 5 we assumed that the first trial solution is always feasible. It is now appropriate to discuss what to do if this is not so. Dantzig's original approach is simple and elegant. This is to start with a trial solution in which all the genuine, or *structural*, variables are non-basic. If all constraints are either less-than-or-equal-to constraints with a non-negative right-hand side or greater-than-or-equal-to constraints with a non-positive right-hand side, then we have a feasible trial solution by using the slack variables as the set of basic variables. If there are equality constraints then we can introduce *artificial variables* representing the differences between the left- and right-hand sides of each equality constraint. We can then choose the sign of the artificial so that its numerical value is non-negative when the structural variables are all zero. Also, if there are inequality constraints that are not satisfied initially, we can first add slack variables to turn them into equality equations and then add artificials.

We now have a tableau with the formal appearance of one with a feasible trial solution without having done any arithmetic. However, this does not represent a genuinely feasible solution unless all the artificials vanish. We therefore set up an *auxiliary objective function* to minimize the sum of the values of the artificials. We treat these artificials as variables required to be non-negative, so the auxiliary objective function will vanish if and only if the corresponding trial solution is genuinely feasible.

The process of minimizing the auxiliary objective function is sometimes called *phase one* of the simplex method. If a zero value is not achieved, the whole problem is *infeasible*. If it is achieved, we turn to *phase two*, which is to maximize x_0 without increasing the auxiliary objective function.

In practice we do not allow artificials to become basic again once they have become non-basic. This may prevent us from achieving the true minimum of the auxiliary objective function, but it cannot stop us finding a feasible solution if one exists. Often the artificials have all become non-basic by the start of phase two, but this need not be so if the trial value of an artificial remains zero. The rules for pivoting therefore have to be extended to say that if X_i is artificial and $\bar{a}_{iq} \neq 0$ one should pivot in this row rather than increase X_{m+q} by a positive amount. These ideas can be illustrated by considering the blending problem illustrated in Table 6.1.1, which is an extended version of the example in Table 5.2.1.

The problem is to maximize x_0 subject to

$$
\begin{aligned}
x_0 + 4x_1 + 5x_2 + 7x_3 + 6x_4 &= 0 \\
x_1 + x_2 + x_3 + x_4 &= 1 \\
5x_2 + 3x_3 + 5x_4 &\geqslant 2 \\
x_2 + 2x_3 + 2x_4 &\geqslant 1 \\
10x_1 \quad\quad + 4x_3 + 2x_4 &= 5 \\
4x_2 + x_3 + x_4 &\leqslant 2
\end{aligned}
$$

Introducing non-negative slack variables to turn all the equations into equality constraints we have

$$
\begin{aligned}
x_0 + 4x_1 + 5x_2 + 7x_3 + 6x_4 &= 0 \\
x_1 + x_2 + x_3 + x_4 &= 1 \\
5x_2 + 3x_3 + 5x_4 - s_2 &= 2 \\
x_2 + 2x_3 + 2x_4 \quad - s_3 &= 1 \\
10x_1 \quad\quad + 4x_3 + 2x_4 &= 5 \\
4x_2 + x_3 + x_4 \quad\quad + s_5 &= 2
\end{aligned}
$$

Introducing artificial variables, the initial tableau reads

$$
\begin{aligned}
x_0 &= 0 - 4x_1 - 5x_2 - 7x_3 - 6x_4 \\
A_1 &= 1 - x_1 - x_2 - x_3 - x_4 \\
A_2 &= 2 \quad\quad - 5x_2 - 3x_3 - 5x_4 + s_2 \\
A_3 &= 1 \quad\quad - x_2 - 2x_3 - 2x_4 \quad + s_3 \\
A_4 &= 5 - 10x_1 \quad\quad - 4x_3 - 2x_4 \\
s_5 &= 2 \quad\quad - 4x_2 - x_3 - x_4 \\
S &= 9 - 11x_1 - 7x_2 - 10x_3 - 10x_4 + s_2 + s_3
\end{aligned}
$$

Table 6.1.1

Quality	Raw material Quality restriction (%)	Sago flour	Meat meal	Enriched skimmed milk	Soya flour
			Amount of quality in raw material (%)		
Protein	$\geqslant 20$	0	50	30	50
Fat	$\geqslant 10$	0	10	20	20
Carbohydrate	$= 50$	100	0	40	20
Fibre and ash	$\leqslant 20$	0	40	10	10

where S is the *sum of infeasibilities* given by $A_1 + A_2 + A_3 + A_4$. We begin by trying to minimize S, so we increase the variable in the expression for S which has the most negative coefficient, i.e. x_1. We increase x_1 until A_4 becomes equal to zero. Therefore the tableau becomes

$$
\begin{aligned}
x_0 &= -2 + \tfrac{4}{10}A_4 - 5x_2 - \tfrac{54}{10}x_3 - \tfrac{52}{10}x_4 \\
A_1 &= \tfrac{5}{10} + \tfrac{1}{10}A_4 - x_2 - \tfrac{6}{10}x_3 - \tfrac{8}{10}x_4 \\
A_2 &= 2 \qquad\quad - 5x_2 - 3x_3 - 5x_4 + s_2 \\
A_1 &= 1 \qquad\quad - x_2 - 2x_3 - 2x_4 \qquad + s_3 \\
x_1 &= \tfrac{5}{10} - \tfrac{1}{10}A_4 \qquad\quad - \tfrac{1}{10}x_3 - \tfrac{2}{10}x_4 \\
s_5 &= 2 \qquad\quad - 4x_2 - x_3 - x_4 \\
S &= \tfrac{35}{10} + \tfrac{11}{10}A_4 - 7x_2 - \tfrac{56}{10}x_3 - \tfrac{78}{10}x_4 + s_2 + s_3
\end{aligned}
$$

We now increase x_4 until A_2 becomes zero:

$$
\begin{aligned}
x_0 &= -\tfrac{204}{50} + \tfrac{20}{50}A_4 + \tfrac{10}{50}x_2 - \tfrac{114}{50}x_3 + \tfrac{52}{50}A_2 - \tfrac{52}{50}s_2 \\
A_1 &= \tfrac{9}{50} + \tfrac{5}{50}A_4 - \tfrac{10}{50}x_2 - \tfrac{6}{50}x_3 + \tfrac{8}{50}A_2 - \tfrac{8}{50}s_2 \\
x_4 &= \tfrac{20}{50} \qquad\quad - x_2 - \tfrac{30}{50}x_3 - \tfrac{10}{50}A_2 + \tfrac{10}{50}s_2 \\
A_3 &= \tfrac{10}{50} \qquad\quad + x_2 - \tfrac{40}{50}x_3 + \tfrac{20}{50}A_2 - \tfrac{20}{50}s_2 + s_3 \\
x_1 &= \tfrac{21}{50} - \tfrac{5}{50}A_4 + \tfrac{10}{50}x_2 - \tfrac{14}{50}x_3 + \tfrac{2}{50}A_2 - \tfrac{2}{50}s_2 \\
s_5 &= \tfrac{80}{50} \qquad\quad - 3x_2 - \tfrac{20}{50}x_3 + \tfrac{10}{50}A_2 - \tfrac{10}{50}s_2 \\
S &= \tfrac{19}{50} + \tfrac{55}{50}A_4 + \tfrac{40}{50}x_2 - \tfrac{46}{50}x_3 + \tfrac{78}{50}A_2 - \tfrac{28}{50}s_2 + s_3
\end{aligned}
$$

Now increase x_3 until A_3 is zero:

$$
\begin{aligned}
x_0 &= -\tfrac{186}{40} + \tfrac{16}{40}A_4 - \tfrac{106}{40}x_2 + \tfrac{114}{40}A_3 - \tfrac{4}{40}A_2 + \tfrac{4}{40}s_2 - \tfrac{114}{40}s_3 \\
A_1 &= \tfrac{6}{40} + \tfrac{4}{40}A_4 - \tfrac{14}{40}x_2 + \tfrac{6}{40}A_3 + \tfrac{4}{40}A_2 - \tfrac{4}{40}s_2 - \tfrac{6}{40}s_3 \\
x_4 &= \tfrac{10}{40} \qquad\quad - \tfrac{70}{40}x_2 + \tfrac{30}{40}A_3 - \tfrac{20}{40}A_2 + \tfrac{20}{40}s_2 - \tfrac{30}{40}s_3 \\
x_3 &= \tfrac{10}{40} \qquad\quad + \tfrac{50}{40}x_2 - \tfrac{50}{40}A_3 + \tfrac{20}{40}A_2 - \tfrac{20}{40}s_2 + \tfrac{50}{40}s_3 \\
x_1 &= \tfrac{14}{40} - \tfrac{4}{40}A_4 - \tfrac{6}{40}x_2 + \tfrac{14}{40}A_3 - \tfrac{4}{40}A_2 + \tfrac{4}{40}s_2 - \tfrac{14}{40}s_3 \\
s_5 &= \tfrac{60}{40} \qquad\quad - \tfrac{140}{40}x_2 + \tfrac{20}{40}A_3 \qquad\qquad - \tfrac{20}{40}s_3 \\
S &= \tfrac{6}{40} + \tfrac{44}{40}A_4 - \tfrac{14}{40}x_2 + \tfrac{46}{40}A_3 + \tfrac{44}{40}A_2 - \tfrac{4}{40}s_2 - \tfrac{6}{40}s_3
\end{aligned}
$$

If we now increase x_2 until x_4 becomes zero, the tableau becomes

$$
\begin{aligned}
x_0 &= -\tfrac{352}{70} + \tfrac{28}{70}A_4 + \tfrac{106}{70}x_4 + \tfrac{120}{70}A_3 + \tfrac{46}{70}A_2 - \tfrac{46}{70}s_2 - \tfrac{120}{70}s_3 \\
A_1 &= \tfrac{7}{70} + \tfrac{7}{70}A_4 + \tfrac{14}{70}x_4 \qquad\qquad + \tfrac{14}{70}A_2 - \tfrac{14}{70}s_2 \\
x_2 &= \tfrac{10}{70} \qquad\quad - \tfrac{40}{70}x_4 + \tfrac{30}{70}A_3 - \tfrac{20}{70}A_2 + \tfrac{20}{70}s_2 - \tfrac{30}{70}s_3 \\
x_3 &= \tfrac{30}{70} \qquad\quad - \tfrac{50}{70}x_4 - \tfrac{50}{70}A_3 + \tfrac{10}{70}A_2 - \tfrac{10}{70}s_2 + \tfrac{50}{70}s_3 \\
x_1 &= \tfrac{23}{70} - \tfrac{7}{70}A_4 + \tfrac{6}{70}x_4 + \tfrac{20}{70}A_3 - \tfrac{4}{70}A_2 + \tfrac{4}{70}s_2 - \tfrac{20}{70}s_3 \\
s_5 &= 1 \qquad\quad + 2x_4 - A_3 + A_2 - s_2 + s_3 \\
S &= \tfrac{7}{70} + \tfrac{77}{70}A_4 + \tfrac{14}{70}x_4 + A_3 + \tfrac{84}{70}A_2 - \tfrac{14}{70}s_2
\end{aligned}
$$

Making s_2 basic and A_1 non-basic,

$$x_0 = -\tfrac{75}{14} + \tfrac{1}{14}A_4 + \tfrac{12}{14}x_4 + \tfrac{24}{14}A_3 \qquad\quad + \tfrac{46}{14}A_1 - \tfrac{24}{14}s_3$$
$$s_2 = \tfrac{7}{14} + \tfrac{7}{14}A_4 + \quad x_4 \qquad\qquad + A_2 - 5A_1$$
$$x_2 = \tfrac{4}{14} + \tfrac{2}{14}A_4 - \tfrac{4}{14}x_4 + \tfrac{6}{14}A_3 \qquad\quad - \tfrac{20}{14}A_1 - \tfrac{6}{14}s_3$$
$$x_3 = \tfrac{5}{14} - \tfrac{1}{14}A_4 - \tfrac{12}{14}x_4 - \tfrac{10}{14}A_3 \qquad\quad + \tfrac{10}{14}A_1 + \tfrac{10}{14}s_3$$
$$x_1 = \tfrac{5}{14} - \tfrac{1}{14}A_4 + \tfrac{2}{14}x_4 + \tfrac{4}{14}A_3 \qquad\quad - \tfrac{4}{14}A_1 - \tfrac{4}{14}s_3$$
$$s_5 = \tfrac{7}{14} - \tfrac{7}{14}A_4 + \quad x_4 - \quad A_3 \qquad\quad + 5A_1 + \quad s_3$$
$$S = \qquad\qquad A_4 \qquad\quad + \quad A_3 + A_2 + \quad A_1$$

The trial solution is now feasible but not optimum, because we are trying to maximize x_0 and there are positive coefficients in the expression for x_0. We cannot increase any of the artificial variables but we can increase x_4 until $x_3 = 0$, obtaining

$$x_0 = -5 \qquad\quad - \quad x_3 + \quad A_3 \qquad\quad + 4A_1 - \quad s_3$$
$$s_2 = \tfrac{11}{12} + \tfrac{5}{12}A_4 - \tfrac{14}{12}x_3 - \tfrac{10}{12}A_3 + A_2 - \tfrac{50}{12}A_1 + \tfrac{10}{12}s_3$$
$$x_2 = \tfrac{2}{12} + \tfrac{2}{12}A_4 + \tfrac{4}{12}x_3 + \tfrac{8}{12}A_3 \qquad\quad - \tfrac{20}{12}A_1 - \tfrac{8}{12}s_3$$
$$x_4 = \tfrac{5}{12} - \tfrac{1}{12}A_4 - \tfrac{14}{12}x_3 - \tfrac{10}{12}A_3 \qquad\quad + \tfrac{10}{12}A_1 + \tfrac{10}{12}s_3$$
$$x_1 = \tfrac{5}{12} - \tfrac{1}{12}A_4 - \tfrac{2}{12}x_3 + \tfrac{2}{12}A_3 \qquad\quad - \tfrac{2}{12}A_1 - \tfrac{2}{12}s_3$$
$$s_5 = \tfrac{11}{12} - \tfrac{7}{12}A_4 - \tfrac{14}{12}x_3 - \tfrac{22}{12}A_3 \qquad\quad + \tfrac{70}{12}A_1 + \tfrac{22}{12}s_3$$

This is the final solution, since x_0 can only be increased by increasing an artificial variable, which is not permitted. The minimum cost is therefore £50 per tonne, obtained by using the raw materials sago flour, meat meal, enriched skimmed milk and soya flour in the ratio 5:2:0:5. The slack variables s_2 and s_5 are both $\tfrac{11}{12}$, indicating that the optimum solution is $\tfrac{11}{12}$ of 10%, or about 9.2%, over the minimum protein requirement and the same amount under the maximum fibre and ash limit.

What about the reduced costs? The reduced cost of x_3 is 1, indicating a unit rate of increase of cost with x_3. However, the reduced costs of the slack and artificial variables have a special significance. They are known as the *shadow prices*, and indicate the rate of change of the objective function with respect to the right-hand sides of the constraints. In the example the shadow prices are the reduced costs of the variables A_1, A_2, A_3, A_4 and s_5, and take the values -4, 0, -1, 0 and 0, respectively. The reduced cost of A_1 is not of much interest, since it would be physically meaningless to change the right-hand side of the first constraint, but the other reduced costs represent the effects of small changes in the other limits. It appears that the only crucial requirement is that for fat. The cost increases at a rate of one unit per unit increase in the fat requirement, i.e. the objective function increases by £10 per tonne for every 10% increase in this right-hand side.

6.2   PRACTICAL METHOD OF FINDING A FIRST FEASIBLE SOLUTION

We now consider a more efficient method of finding a first feasible solution. Although it is convenient to use artificials for equality constraints, it is clumsy and unnecessary to use these as well as slacks on inequality constraints. The artificials in these cases can be avoided if we allow the slacks on these constraints to take negative values. We now need to devise a systematic reliable method of proceeding when some of the $\bar{a}_{i0}$ are negative.

One such method, the *composite simplex algorithm* described by Wolfe (1965), is to define the sum of infeasibilities as the sum of the absolute values of the artificials plus that of the absolute values of all basic variables in the current solution that have negative values. In other words, if we are considering the tableau

$$x_0 = \bar{a}_{00} - \sum_{j=1}^{n-m} \bar{a}_{0j} X_{m+j}$$

$$X_i = \bar{a}_{i0} - \sum_{j=1}^{n-m} \bar{a}_{ij} X_{m+j} \qquad (i = 1, \ldots, m) \qquad (6.2.1)$$

the sum of infeasibilities is

$$S = \sum_{\substack{i=1 \\ i \in A}}^{m} |X_i| + \sum_{\substack{i=1 \\ i \in N}}^{m} |X_i|$$

where $i \in A$ if X_i is an artificial variable and $\bar{a}_{i0} > 0$ and $i \in N$ if X_i is a basic variable with $\bar{a}_{i0}$ less than zero. Note that if X_i is an artificial it will always be non-negative and basic, since we will remove it from the tableau as soon as it becomes non-basic. Thus the sum of infeasibilities S is given by

$$S = \sum_{\substack{i=1 \\ i \in A}}^{m} X_i - \sum_{\substack{i=1 \\ i \in N}}^{m} X_i$$

Using the tableau equations (6.2.1) we can rewrite this sum in terms of the non-basic variables:

$$S = \sum_{\substack{i=1 \\ i \in A}}^{m} \bar{a}_{i0} - \sum_{\substack{i=1 \\ i \in N}}^{m} \bar{a}_{i0} + \sum_{j=1}^{n-m} d_j X_{m+j}$$

where

$$d_j = - \sum_{\substack{i=1 \\ i \in A}}^{m} \bar{a}_{ij} + \sum_{\substack{i=1 \\ i \in N}}^{m} \bar{a}_{ij}$$

The quantities d_j are known as the *reduced costs* of the non-basic variables X_{m+j} and can be rewritten as

$$d_j = \sum_{i=1}^{m} c_i \bar{a}_{ij}$$

where

$$c_i = \begin{cases} -1 & \text{if } i \in A \\ +1 & \text{if } i \in N \\ 0 & \text{otherwise} \end{cases}$$

If all the d_j's are positive and the sum of infeasibilities S still has a value greater than zero, the problem is *infeasible*. However, if the reduced cost d_q is negative, we can increase the non-basic variable X_{m+q} and reduce S, which is then a piecewise linear function of X_{m+q}. Initially S is reduced at a rate of $-d_q$, but this rate decreases whenever the value of a basic variable changes sign. It is therefore reasonably straightforward to choose the increase in X_{m+q} to maximize the decrease in the sum of infeasibilities, and then to pivot in the row defining the basic variable that is driven to zero.

In practice it is more efficient to pay some attention to the ultimate objective function from the beginning, and to define d_j by

$$d_j = W\bar{a}_{0j} + \sum_{i=1}^{m} c_i \bar{a}_{ij}$$

for some small positive weight W. However, we must then reduce W—perhaps to zero—if we reach an infeasible solution with no non-basic variable having a negative reduced cost d_j. This method applied to the previous example, with $W = 0.1$, works as follows:

$$
\begin{aligned}
x_0 &= 0 - 4x_1 - 5x_2 - 7x_3 - 6x_4 \\
A_1 &= 1 - x_1 - x_2 - x_3 - x_4 \\
s_2 &= -2 + 5x_2 + 3x_3 + 5x_4 \\
s_3 &= -1 + x_2 + 2x_3 + 2x_4 \\
A_4 &= 5 - 10x_1 - 4x_3 - 2x_4 \\
s_5 &= 2 - 4x_2 - x_3 - x_4
\end{aligned}
$$

Therefore since A_1 and A_4 are artificials and s_2 and s_3 are negative basic variables, the values of $c_1, \ldots, c_5$ are $-1, 1, 1, -1, 0$ and the values of the reduced costs $d_1, \ldots, d_4$ are $-10.6, -6.5, -9.3$ and -9.4, respectively. So we choose to increase the non-basic variable x_1 since d_1 is the most negative reduced cost. The first basic variable to reach zero is A_4, so we make it non-basic. We then have

$$
\begin{aligned}
x_0 &= -2 + \tfrac{4}{10}A_4 - 5x_2 - \tfrac{54}{10}x_3 - \tfrac{52}{10}x_4 \\
A_1 &= \tfrac{5}{10} + \tfrac{1}{10}A_4 - x_2 - \tfrac{6}{10}x_3 - \tfrac{8}{10}x_4 \\
s_2 &= -2 + 5x_2 + 3x_3 + 5x_4 \\
s_3 &= -1 + x_2 + 2x_3 + 2x_4 \\
x_1 &= \tfrac{5}{10} - \tfrac{1}{10}A_4 - \tfrac{4}{10}x_3 - \tfrac{2}{10}x_4 \\
s_5 &= 2 - 4x_2 - x_3 - x_4
\end{aligned}
$$

The reduced costs of the non-artificial, non-basic variables are now -6.5, -5.06 and -7.28, so we now increase x_4. When $x_4 = 0.4$, $s_2 = 0$ and the reduced cost for x_4 increases by 5 to -2.28. However, it is still negative, so we can increase x_4 further. When $x_4 = 0.5$, $s_3 = 0$ and the reduced cost increases by 2 to -0.28. It is, however, still negative, and all that has happened is that two negative basic variables have become positive, so we increase x_4 even further. However, when $x_4 = 0.625$, the artificial variable A_1 becomes zero, so we make it non-basic.

We then have

$$
\begin{aligned}
x_0 &= -\tfrac{42}{8} - \tfrac{2}{8}A_4 + \tfrac{12}{8}x_2 - \tfrac{12}{8}x_3 + \tfrac{52}{8}A_1 \\
x_4 &= \tfrac{5}{8} + \tfrac{1}{8}A_4 - \tfrac{10}{8}x_2 - \tfrac{6}{8}x_3 - \tfrac{10}{8}A_1 \\
s_2 &= \tfrac{9}{8} + \tfrac{5}{8}A_4 - \tfrac{10}{8}x_2 - \tfrac{6}{8}x_3 - \tfrac{50}{8}A_1 \\
s_3 &= \tfrac{2}{8} + \tfrac{2}{8}A_4 - \tfrac{12}{8}x_2 + \tfrac{4}{8}x_3 - \tfrac{20}{8}A_1 \\
x_1 &= \tfrac{3}{8} - \tfrac{1}{8}A_4 + \tfrac{2}{8}x_2 - \tfrac{2}{8}x_3 + \tfrac{2}{8}A_1 \\
s_5 &= \tfrac{11}{8} - \tfrac{1}{8}A_4 - \tfrac{22}{8}x_2 - \tfrac{2}{8}x_3 + \tfrac{10}{8}A_1
\end{aligned}
$$

We now have a feasible trial solution, because all the artificial variables have been made non-basic and the basic variables are non-negative as required. We can now proceed with phase two in the normal way and increase x_2 because we are trying to maximize x_0. The first basic variable to become zero on increasing x_2 is s_3, so we make s_3 non-basic:

$$
\begin{aligned}
x_0 &= -5 \qquad\quad - \quad s_3 - \quad x_3 + \ 4A_1 \\
x_4 &= \tfrac{5}{12} - \tfrac{1}{12}A_4 + \tfrac{10}{12}s_3 - \tfrac{14}{12}x_3 + \tfrac{10}{12}A_1 \\
s_2 &= \tfrac{11}{12} + \tfrac{5}{12}A_4 + \tfrac{10}{12}s_3 - \tfrac{14}{12}x_3 - \tfrac{50}{12}A_1 \\
x_2 &= \tfrac{2}{12} + \tfrac{2}{12}A_4 - \tfrac{8}{12}s_3 + \tfrac{4}{12}x_3 - \tfrac{20}{12}A_1 \\
x_1 &= \tfrac{5}{12} - \tfrac{1}{12}A_4 - \tfrac{2}{12}s_3 - \tfrac{2}{12}x_3 - \tfrac{2}{12}A_1 \\
s_5 &= \tfrac{11}{12} - \tfrac{7}{12}A_4 + \tfrac{22}{12}s_3 - \tfrac{14}{12}x_3 + \tfrac{70}{12}A_1
\end{aligned}
$$

We have achieved the same optimum solution in half the number of steps needed by the method in Section 6.1.

6.3   UPPER AND LOWER BOUNDS

So far, we have assumed that the only bounds on the decision variables are lower ones of zero (and upper bounds of infinity). Often, however, variables have both non-zero lower and finite upper bounds. In other words,

$$ L_j \leqslant x_j \leqslant U_j \qquad (j = 1, \ldots, n) $$

Lower bounds are easily dealt with using a transformation of the form $\bar{x}_j = x_j - L_j$, so that the problem is defined in terms of the new decision variables $\bar{x}_j$ which have the bounds

$$ 0 \leqslant \bar{x}_j \leqslant \bar{U}_j \qquad (j = 1, \ldots, n) $$

where $\bar{U}_j = U_j - L_j$.

How do we cope with upper bounds? One obvious approach is to treat them as explicit constraints, introducing slack variables for each of the upper bounds. So to solve the simple problem

maximize $\qquad\qquad 5x_1 + 2x_2$

subject to

$$
\begin{aligned}
x_1 + \ x_2 &\leqslant 5 \\
-x_1 + 2x_2 &\leqslant 6 \\
0 \leqslant \ x_1 &\leqslant 3
\end{aligned}
$$

we would have the initial tableau

$$x_0 = \quad 5x_1 + 2x_2$$
$$s_1 = 5 - \quad x_1 - \quad x_2$$
$$s_2 = 6 + \quad x_1 - 2x_2$$
$$s_4 = 3 - \quad x_1$$

However, treating upper bounds as explicit constraints means that we have to add more constraints and slacks to the problem and therefore increase the number of rows and columns in the tableau. This can dramatically add to the amount of computer time needed to solve the problem, particularly if the problem is a large one with many bounded decision variables.

Dantzig (1955) described special facilities for simple upper bounds on individual variables and in the mid-1960s these became a standard feature of computer programs for solving linear programming problems. These facilities are based on the fact that either the structural variable x_j, or the slack on this constraint, or both, must be basic at any time. Therefore the row of the tableau defining one of these basic variables can then always be deduced from the rest of the tableau. Thus this row, and the corresponding basic variable, can be treated implicitly instead of explicitly. In practice, it is more convenient to work with the negative of the slack. This is a non-positive *excess variable*, representing $x_j - U_j$. When the excess variable is non-basic, x_j is said to be out of the basis at its upper bound. For example, consider solving the earlier simple linear programming problem:

$$x_0 = \quad 5x_1 + 2x_2$$
$$s_1 = 5 - \quad x_1 - \quad x_2$$
$$s_2 = 6 + \quad x_1 - 2x_2$$

We begin by increasing x_1 because it has the largest positive coefficient in the expression for x_0. The tableau suggests that we should increase it to the value 5, when s_1 becomes zero. However, x_1 has an upper bound of 3, so we increase it to this upper bound and rewrite the tableau in terms of the non-basic excess variable $x_1' = x_1 - 3$ to indicate that x_1 is now at its upper bound and cannot be increased any further.

$$x_0 = 15 + 5x_1' + 2x_2$$
$$s_1 = \quad 2 - \quad x_1' - \quad x_2$$
$$s_2 = \quad 9 + \quad x_1' - 2x_2$$

Next, we increase the variable x_2 until s_1 becomes zero:

$$x_0 = 19 + 3x_1' - 2s_1$$
$$x_2 = \quad 2 - \quad x_1' - \quad s_1$$
$$s_2 = \quad 5 - \quad x_1' + 2s_1$$

This is the optimum solution because the coefficient of s_1 is negative and the variable x_1 is at its upper bound. Thus the optimum solution is $x_1 = 3$, $x_2 = 2$.

Other types of constraints can be treated in a similar way. Dantzig and Van

Slyke (1967) defined a constraint of the form

$$\sum_j x_{jk} = b_k$$

as a *generalized upper bound*, and Schrage (1975) defined a constraint of the form

$$x_j - x_k \leqslant 0$$

as a *variable upper bound*. Both types of constraint arise frequently in applications and both have been implemented in commercial computer programs for solving linear programming problems. Neither, however, has proved useful enough to become an established feature.

6.4   EXPLOITING SPARSITY

In nearly all practical linear programming problems a typical variable only occurs in about six or fewer constraints. Thus large problems are nearly always very *sparse*. General computer programs for solving linear programming problems must be applicable to large problems, so they must use algorithms which exploit this sparseness efficiently, even if it makes them slower than special-purpose programs on small dense problems. The simplex method has retained its central position in computer programs for solving linear programming problems because successive implementations of the method have been able to exploit sparseness increasingly more efficiently. We will now concentrate on how the simplex method is implemented in computer programs and see how sparseness is exploited.

A general linear programming problem can be written in matrix notation as

maximize x_0

subject to $\mathbf{Ax} = \mathbf{b}$ (6.4.1)

where $\mathbf{A}$ is an $m \times n$ matrix (including the objective function) and where $\mathbf{x}$ is an $n \times 1$ column vector (including the dummy variable x_0). In principle, the simplex method solves equation (6.4.1) for the basic variables, say $\mathbf{x_B}$, in terms of the non-basic variables $\mathbf{x_N}$. So assuming that the matrix $\mathbf{A}$ has been reordered so that the first m columns correspond to the basic variables, we can write equation (6.4.1) in the form

$$\mathbf{Bx_B} + \mathbf{Nx_N} = \mathbf{b}$$ (6.4.2)

where $\mathbf{B}$ is the $m \times m$ square matrix formed from the columns of $\mathbf{A}$ associated with the basic variables and $\mathbf{N}$ denotes the remaining columns of $\mathbf{A}$, i.e. the coefficients of the non-basic variables. Pre-multiplying equation (6.4.2) by $\mathbf{B}^{-1}$, it follows that

$$\mathbf{x_B} = \boldsymbol{\beta} - \mathbf{B}^{-1}\mathbf{Nx_N}$$ (6.4.3)

where $\boldsymbol{\beta} = \mathbf{B}^{-1}\mathbf{b}$. The matrix $\mathbf{B}^{-1}\mathbf{N}$ defines the coefficients of the non-basic variables in the tableau for the original or straight simplex method. The column vector $\boldsymbol{\beta}$ defines the value of the objective function and the values of the basic variables, i.e. the $\bar{a}_{i0}$.

In the original simplex method the whole tableau is computed and updated explicitly at each iteration, even though very few of the coefficients are ever used in a single iteration. Specifically, those which are used are:

(1) All the coefficients in the objective function, to select the new pivotal column (normally the one with the most negative reduced cost);
(2) The right-hand side and elements in the pivotal column, to select the new pivotal row;
(3) The other elements in the pivotal row, to update the expression for the objective function.

The remaining columns are updated simply because they may be needed as pivotal columns in a subsequent iteration.

However, equation (6.4.3) shows that there is no need to calculate the whole tableau explicitly at each iteration. Instead the relevant features of the tableau can be computed from the original matrix $\mathbf{A}$ if we maintain some representation of the inverse matrix $\mathbf{B}^{-1}$. Thus we need to store the original matrix $\mathbf{A}$ (which is stored as a sparse matrix by columns), the list of basic variables $x_0, X_1, \ldots, X_m$, the right-hand side vector $\boldsymbol{\beta}$ and some expression for $\mathbf{B}^{-1}$. This is known as the *revised simplex method*, due to Dantzig *et al.* (1955). When the basis changes, there is no need to invert the new matrix $\mathbf{B}$ again. Instead the inverse matrix $\mathbf{B}^{-1}$ can be updated directly from one iteration to the next in the same way that the tableau is updated in the original simplex method.

The basic steps of the revised simplex method (sometimes known as the *inverse matrix method*) are as follows.

Step 1

Produce a pricing vector $\boldsymbol{\pi}$ such that the elements of the product $\boldsymbol{\pi}\mathbf{N}$ are the reduced costs of the non-basic variables. The form of this step depends on whether the current trial solution is feasible or not. If it is, the reduced costs are the top row of the matrix $\mathbf{B}^{-1}\mathbf{N}$. This can be expressed in matrix notation as $\mathbf{cB}^{-1}\mathbf{N}$, where $\mathbf{c} = (1, 0, \ldots, 0)^T$. Therefore in this case the vector $\boldsymbol{\pi} = \mathbf{cB}^{-1}$, and if $\mathbf{B}^{-1}$ is stored explicitly, this operation simply involves picking out its top row.

If the current trial solution is not feasible, we must form reduced costs that measure the rates of increase in the sum of infeasibilities, so we use Wolfe's procedure with the row vector $\mathbf{c}$ defined as described in Section 6.2. In other words, $c_i = -1$ if the basic variable is artificial and $\bar{a}_{i0} > 0$; $c_i = +1$ if $\bar{a}_{i0} < 0$ (and $i > 0$); $c_i = 0$ otherwise.

Step 2

Price out the columns of $\mathbf{N}$ to find a variable to remove from the set of non-basic variables. This can be done by forming every element of $\boldsymbol{\pi}\mathbf{N}$ and choosing the most negative. However, the choice of column can be improved by taking into account the size of the entries in the columns using *scale factors* (e.g. Harris, 1973) for the reduced costs. Looking at every column of $\pi\mathbf{N}$ can be a slow process, so some implementations of the revised simplex method only scan a subset of the columns of $\mathbf{N}$ during any one iteration.

Step 3

If X_{m+q} is the non-basic variable chosen in step 2, form the updated column $\boldsymbol{\alpha}$ of coefficients $\bar{a}_{iq}$ of this variable by multiplying the appropriate column of $\mathbf{N}$ by $\mathbf{B}^{-1}$.

Step 4

Perform a ratio test between the elements of $\boldsymbol{\alpha}$ and the elements of $\boldsymbol{\beta}$ to determine the pivot, and the variable to be made non-basic.

Step 5

Update the set of basic variables, the vector $\boldsymbol{\beta}$ and the inverse $\mathbf{B}^{-1}$. The last operation involves pre-multiplying $\mathbf{B}^{-1}$ by an elementary column transformation. This is a unit matrix, except for the pth column, which has $1/\bar{a}_{pq}$ as its diagonal entry and $-\bar{a}_{iq}/\bar{a}_{pq}$ in the ith row for $i \neq p$. The columns other than the pth play such a minor role that these matrices are sometimes called vectors, specifically *eta-vectors*.

   If the matrix $\mathbf{B}^{-1}$ is stored explicitly, the resulting algorithm is called the *explicit inverse* algorithm. The *product form* algorithm, however, due to Dantzig and Orchard Hays (1954) is a better algorithm because it exploits the sparseness of $\mathbf{B}$. In this algorithm, $\mathbf{B}^{-1}$ is stored as a sequence of *elementary column transformations*, say $\mathbf{T}_1, \mathbf{T}_2, \ldots$. Step 1 of the revised simplex method then involves post-multiplying the row vector $\mathbf{c}$ by each of these elementary matrices successively, in the opposite order to that in which they were generated, e.g. $\mathbf{cT}_4\mathbf{T}_3\mathbf{T}_2\mathbf{T}_1$. Step 3 involves pre-multiplying the appropriate column of $\mathbf{N}$ by the elementary matrices successively in the order in which they were generated.

   After a while, the list of elementary matrices becomes undesirably long and rounding-off errors accumulate, so that the product no longer accurately represents the inverse. Furthermore, in reaching the current basis a variable may have been introduced into the basis and removed from it several times. However, the only information which is important for the next iteration is the list of variables which are currently basic and those which are non-basic. Therefore the complete list of elementary matrices is not strictly needed.

   We thus produce a shorter list of elementary matrices by a process known as *re-inversion*. Given the list of variables which are currently basic we can form the current basis matrix $\mathbf{B}$ from the appropriate columns of the original matrix $\mathbf{A}$. We then calculate at most m elementary matrices, say $\mathbf{R}_1, \mathbf{R}_2, \ldots, \mathbf{R}_m$, representing the pivotal operations needed to reduce this matrix $\mathbf{B}$ to the unit matrix I. In other words,

$$R_m \cdots R_3 R_2 R_1 B = I$$

These elementary matrices will then represent the inverse matrix B^{-1}. During this re-inversion process we have considerable freedom in the choice of the pivotal rows and columns, i.e. the order in which we choose to eliminate non-zero elements from the matrix B as we reduce it to the unit matrix I.

Many improvements were made during the 1960s in the inversion routine, which is called every 100 iterations or so. The problem is to choose the best order to pivot on the columns during the re-inversion process, so that the product form representation of the inverse is as sparse as possible. It is easy to show that pivoting on the diagonal elements of a lower triangular matrix, in order, produces elementary matrices which contain no more non-zeroes than the original matrix itself. Thus it is worth permuting the rows and columns of B (before calculating the elementary matrices during re-inversion) so that they are as nearly in lower triangular form as possible (see Hellerman and Rarick, 1971).

Around 1970 it became generally recognized that sparseness is better exploited by finding L and U such that

$$B = LU$$

where L is a lower triangular matrix and U is an upper triangular one. A product form representation of B^{-1} can then be formed without creating any further non-zeroes by pivoting on the diagonal elements of L, and then on the diagonal elements of U in reverse order. The essence of this idea is contained in Markowitz (1957).

7

Duality and Parametric Programming

7.1 INTRODUCTION TO DUALITY

Since the principle of duality applies to both non-linear and linear programming we will begin by considering the general non-linear case before specializing to linear programming later. Let us therefore start by considering the following general non-linear programming problem with n non-negative variables and m constraints:

minimize $f(\mathbf{x})$
subject to

$$g_i(\mathbf{x}) \leqslant b_i \qquad (i = 1, \ldots, m)$$
$$x_j \geqslant 0 \qquad (j = 1, \ldots, n)$$

We will call this the *primal* problem. By introducing slack variables y_i, for $i = 1, \ldots, m$, and s_j, for $j = 1, \ldots, n$, we can rewrite the constraints as equality constraints:

$$g_i(\mathbf{x}) + y_i = b_i \qquad (i = 1, \ldots, m)$$
$$x_j - s_j = 0 \qquad (j = 1, \ldots, n)$$

Since each of these constraints must be satisfied, the problem is unaffected if we replace the objective function $f(\mathbf{x})$ by

$$f(\mathbf{x}) + \sum_{i=1}^{m} \pi_i(g_i(\mathbf{x}) + y_i - b_i) - \sum_{j=1}^{n} \delta_j(x_j - s_j) \qquad (7.1.1)$$

where the quantities $\pi_1, \ldots, \pi_m$ and $\delta_1, \ldots, \delta_n$ can take any values. This new objective function is known as a *Lagrangian* function and the π_i and δ_j are called *Lagrange multipliers*. If we now remove the equality constraints, the problem becomes one of finding non-negative y_i and s_j to minimize equation (7.1.1), and the minimum of this new problem is certainly not greater than that of the original problem.

If any π_i is negative we can make the objective function arbitrarily large and negative by giving the corresponding y_i an arbitrarily large value. Similarly, if any δ_j is negative, the objective function can be made arbitrarily large and negative by making s_j arbitrarily large. If, however, all $\pi_i \geqslant 0$ and all $\delta_j \geqslant 0$, then the optimization with respect to y_i and s_j simply requires us to set $\pi_i y_i = 0$ and $\delta_j s_j = 0$ for all i and j. Thus we have the relaxed problem

$$\text{minimize} \qquad f(\mathbf{x}) + \sum_{i=1}^{m} \pi_i(g_i(\mathbf{x}) - b_i) - \sum_{j=1}^{n} \delta_j x_j$$

This relaxed problem may be easier to solve than the primal problem, but this is not much use if it is quite unrealistic. Therefore let us try to make it as realistic as possible by choosing non-negative π_i and δ_j to maximize

$$\min_{\mathbf{x}} \left[f(\mathbf{x}) + \sum_{i=1}^{m} \pi_i(g_i(\mathbf{x}) - b_i) - \sum_{j=1}^{n} \delta_j x_j \right]$$

To make further progress, we now assume that the functions $f(\mathbf{x})$ and $g_i(\mathbf{x})$ $(i = 1, \ldots, m)$ are convex and differentiable. The conditions for a minimum with respect to $\mathbf{x}$ are then

$$\frac{\partial f}{\partial x_j} + \sum_{i=1}^{m} \pi_i \frac{\partial g_i}{\partial x_j} - \delta_j = 0 \qquad (j = 1, \ldots, n)$$

so that we can use these equations to eliminate δ_j from the objective function. We then have the problem

$$\max_{\pi \mathbf{x}} \left[\left(f(\mathbf{x}) - \sum_{j=1}^{n} x_j \frac{\partial f}{\partial x_j} \right) + \sum_{i=1}^{m} \pi_i \left(g_i(\mathbf{x}) - \sum_{j=1}^{n} x_j \frac{\partial g_i}{\partial x_j} - b_i \right) \right]$$

subject to

$$\frac{\partial f}{\partial x_j} + \sum_{i=1}^{m} \pi_i \frac{\partial g_i}{\partial x_j} \geqslant 0 \qquad (j = 1, \ldots, n)$$

We have seen that this maximum is always less than or equal to the minimum value of the objective function in the primal problem. On the other hand, Kuhn and Tucker (1951) showed that if the functions are all convex, this maximum is equal to the minimum in the primal problem.

We can now specialize to linear programming. If

$$f(\mathbf{x}) = - \sum_{j=1}^{n} a_{0j} x_j \quad \text{and} \quad g_i(\mathbf{x}) = \sum_{j=1}^{n} a_{ij} x_j$$

we see that the primal problem becomes

$$\text{maximize} \qquad \sum_{j=1}^{n} a_{0j} x_j$$

$$\text{subject to} \qquad \sum_{j=1}^{n} a_{ij} x_j \leqslant b_i \qquad (i = 1, \ldots, m)$$

$$x_j \geqslant 0 \qquad (j = 1, \ldots, n)$$

and the relaxed problem reduces to

minimize $\qquad\qquad \sum_{i=1}^{m} \pi_i b_i$

subject to $\qquad\qquad \sum_{i=1}^{m} \pi_i a_{ij} \geq a_{0j} \qquad (j = 1, \ldots, n)$

$$\pi_i \geq 0 \qquad (i = 1, \ldots, m)$$

There is a good deal of symmetry about this, and if we relax the relaxed problem in the same way we obtain the primal problem again. The relaxed problem is therefore known as the *dual*.

Duality in linear programming can also be approached algorithmically. The formulae for manipulating the tableau

$$x_0 = \bar{a}_{00} - \sum_{j=1}^{n-m} \bar{a}_{0j} X_{m+j}$$

$$X_i = \bar{a}_{i0} - \sum_{j=1}^{n-m} \bar{a}_{ij} X_{m+j} \qquad (i = 1, \ldots, m)$$

also apply to the tableau

$$y_0 = \bar{a}_{00} + \sum_{i=1}^{m} \bar{a}_{i0} Y_i$$

$$Y_{m+j} = \bar{a}_{0j} + \sum_{i=1}^{m} \bar{a}_{ij} Y_i \qquad (j = 1, \ldots, n - m)$$

If the $\bar{a}_{i0}$ and $\bar{a}_{0j}$ are all non-negative, then we have maximized x_0 subject to the constraints that the X_j are all non-negative, and we have equally minimized y_0 subject to the constraints that the Y_j are all non-negative. Furthermore, the optimum objective function values are equal, and if a variable is basic in the optimum solution to the primal with a value $\bar{a}_{i0}$ the corresponding variable in the dual problem is non-basic with a reduced cost $\bar{a}_{i0}$, and vice versa. However, if either problem is unbounded, then the other is infeasible, although it is possible for both primal and dual to be infeasible.

7.2 APPLICATIONS OF DUALITY

One can sometimes exploit duality computationally. The dual to a problem with n inequality constraints in m variables is a problem with m constraints in n non-negative variables. Thus it may be easier to solve the dual if n is large in comparison with m.

An interesting problem in which both primal and dual problems are relevant is in *zero-sum two-person matrix games*. A game can be represented as a situation in which player A has to choose one of m strategies and player B one of n strategies, without either player having any knowledge of the other's choice. If player A chooses strategy i and player B strategy j then A receives a *payoff* of a_{ij} units from B, where this payoff may be positive, zero or negative. In particular, the losses of one player are the gains of the other, and vice versa. The name *matrix game* arises because the possible payoffs, a_{ij}, to player A from B can be represented by the matrix

$$\begin{pmatrix} a_{11}\, a_{12} \ldots a_{1n} \\ a_{21}\, a_{22} \ldots a_{2n} \\ .\qquad .\quad \ldots\quad . \\ a_{m1}\, a_{m2} \ldots a_{mn} \end{pmatrix}$$

The game may be such that if either player repeatedly uses a single strategy, then the other can exploit this rigidity. Therefore we introduce the concept of a *mixed strategy* and say that player A may choose non-negative numbers p_i $(i = 1,\ldots,m)$ summing to unity, representing the probability that he chooses strategy i. He may then wish to maximize his average payoff v against the most awkward strategy from player B. This is called the *minimax* principle. Thus player A's problem can be formulated mathematically as the linear programming problem:

maximize v subject to

$$\sum_{i=1}^{m} p_i = 1$$

$$\sum_{i=1}^{m} a_{ij}p_i - v \geqslant 0 \qquad (j = 1,\ldots,n)$$

$$p_i \geqslant 0 \qquad (i = 1,\ldots,m)$$

Similarly, player B may consider choosing probabilities q_j to minimize w, his average loss to player A, subject to the constraints

$$\sum_{j=1}^{n} q_j = 1$$

$$\sum_{j=1}^{n} a_{ij}q_j - w \leqslant 0 \qquad (i = 1,\ldots,m)$$

$$q_j \geqslant 0 \qquad (j = 1,\ldots,n)$$

These problems are in fact dual problems, since v and w are not restricted to non-negative values. This proves that the optimum values of v and w are equal, so the pessimistic and cautious approach by the players has not lost any advantage that they could reasonably have expected to gain against such a skilful opponent.

Another class of problem for which dualization can help is that of choosing parameters or variables to minimize the sum of the absolute values of the discrepancies between observed and fitted values. This problem can be written as

choose $x_1,\ldots,x_n$ to minimize

$$\sum_{i=1}^{m} \left| a_{i0} - \sum_{j=1}^{n} a_{ij}x_j \right|$$

which in turn can be written as a linear programming problem with m constraints and $n + 2m$ non-negative variables:

minimize $\qquad \sum_{i=1}^{m} z_i + \sum_{i=1}^{m} z_{-i}$

subject to $\qquad \sum_{j=1}^{n} a_{ij}x_j + z_i - z_{-i} = a_{i0} \qquad (i = 1, \ldots, m)$

$$z_i \geqslant 0 \quad \text{and} \quad z_{-i} \geqslant 0 \qquad (i = 1, \ldots, m)$$

The dual of this problem is

minimize $\qquad \sum_{i=1}^{m} \pi_i a_{i0}$

subject to $\qquad \sum_{i=1}^{m} \pi_i a_{ij} = 0 \qquad (j = 1, \ldots, n)$

$$\pi_i \geqslant -1 \qquad (i = 1, \ldots, m)$$

$$-\pi_i \geqslant -1 \qquad (i = 1, \ldots, m)$$

The last two sets of constraints reduce to

$$-1 \leqslant \pi_i \leqslant 1 \qquad (i = 1, \ldots, m)$$

so that the dual problem has n constraints and m *bounded* variables, which is a more compact problem. This is the best way to solve the problem using standard methods, but it is possible to devise special methods that work even better. Davies (1967) first suggested defining artificial variables

$$A_i = a_{i0} - \sum_{j=1}^{n} a_{ij}x_j$$

and finding the true minimum of

$$\sum_{i=1}^{m} |A_i|$$

This requires an extension of the normal phase-one procedure of the simplex method, since one may need to either increase or decrease a non-basic artificial.

7.3 DUAL SIMPLEX METHOD

In Chapter 6 we studied the (primal) simplex method for manipulating tableaux to solve linear programming problems. However, this is not the only approach, and other rules for choosing pivotal rows and columns may be used, provided that they make irreversible progress towards an optimum solution in some sense.

One approach is to go through the motions of applying the simplex method to the dual problem, but interpreting each step in terms of the original problem. This is the method, due to Lemke (1954), which is commonly known as the *dual simplex* method, and we will now study it in more detail.

Suppose that we somehow find a tableau

$$x_0 = \bar{a}_{00} - \sum_{j=1}^{n-m} \bar{a}_{0j} X_{m+j}$$

$$X_i = \bar{a}_{i0} - \sum_{j=1}^{n-m} \bar{a}_{ij} X_{m+j} \qquad (i = 1, \ldots, m)$$

in which all $\bar{a}_{0j} \geqslant 0$. This is known as a *dual feasible* tableau, because the corresponding trial solution to the dual problem is feasible. Then if all the $\bar{a}_{i0} \geqslant 0$ we have an optimum solution. If not, we choose some p such that $\bar{a}_{p0} < 0$ and select this variable X_p to leave the basis. To find the variable to enter the basis in its place, we calculate the ratios

$$\frac{\bar{a}_{0j}}{|\bar{a}_{pj}|}$$

for all j such that $\bar{a}_{pj} < 0$. If no j exists with $\bar{a}_{pj} < 0$, then the problem is infeasible. Otherwise we choose the variable X_q, where q is the argument with the smallest ratio.

This approach can be justified from first principles as a cutting plane method. That is, we have the optimal solution to a problem whose feasible region is wider than that of the problem we really want to solve. We then find a constraint, or cutting plane, that excludes part of this wider region, including the current trial solution, without excluding any genuinely feasible point.

There remains the problem of getting a first dual feasible solution in the dual simplex method. Suppose that some of the $\bar{a}_{0j}$ are negative, and let $\bar{a}_{0q}$ be the most negative. Now introduce a new constraint into the problem

$$x_{n+1} = \Omega + \sum_{j=1}^{n-m} (-X_{m+j})$$

where Ω is some arbitrarily large number. This constraint states that the sum of the non-basic variables must be less than or equal to Ω and does not cut out any finite solution to the true problem. We then make X_{m+q} basic instead of x_{n+1}. The problem will then be dual feasible, though the tableau will have an extra column in that some of the basic variables will have a term in Ω as well as a genuine constant term. If the problem has an unbounded solution, then these Ω's will remain in the problem. Otherwise, x_{n+1} will eventually become basic, and it can be dropped from the problem, together with all the Ω's. Let us illustrate this with an artificial small example:

maximize $\qquad\qquad x_0 = \qquad 3x_1 - \ x_2 + 2x_3$

subject to $\qquad\qquad x_4 = \quad 6 - 2x_1 - \ x_2 - \ x_3$

$\qquad\qquad\qquad\qquad x_5 = -8 \qquad\quad + 2x_2 + \ x_3$

The trial solution is dual infeasible because x_1 and x_3 have positive coefficients in the objective function. Therefore we introduce the constraint

$$x_6 = \Omega - x_1 - x_2 - x_3$$

and pivot between x_1 and x_6. We then have the tableau

$$
\begin{aligned}
x_0 &= 3\Omega - 3x_6 - 4x_2 - x_3 \\
x_4 &= -2\Omega + 6 + 2x_6 + x_2 + x_3 \\
x_5 &= - 8 + 2x_2 + x_3 \\
x_1 &= \Omega - x_6 - x_2 - x_3
\end{aligned}
$$

Now all the non-basic variables have negative coefficients, so the trial solution is dual feasible. However, the solution is not optimal because x_4 and x_5 have negative trial values. Thus we make x_4 non-basic (since it has the most negative value) and pivot between x_4 and x_3, giving

$$
\begin{aligned}
x_0 &= \Omega + 6 - x_6 - 3x_2 - x_4 \\
x_3 &= 2\Omega - 6 - 2x_6 - x_2 + x_4 \\
x_5 &= 2\Omega - 14 - 2x_6 + x_2 + x_4 \\
x_1 &= -\Omega + 6 + x_6 - x_4
\end{aligned}
$$

Now x_1 is negative, so we make it non-basic. The variable to be made basic must be x_6 if the solution is to remain dual feasible, so we write

$$
\begin{aligned}
x_0 &= 12 - x_1 - 3x_2 - 2x_4 \\
x_3 &= 6 - 2x_1 - x_2 - x_4 \\
x_5 &= - 2 - 2x_1 + x_2 - x_4 \\
x_6 &= \Omega - 6 + x_1 + x_4
\end{aligned}
$$

We now have a finite dual feasible solution, so we drop the variables x_6 and Ω. Since x_5 is still negative the solution is not optimal, and we must pivot between x_5 and x_2. We then have

$$
\begin{aligned}
x_0 &= 6 - 7x_1 - 3x_5 - 5x_4 \\
x_3 &= 4 - 4x_1 - x_5 - 2x_4 \\
x_2 &= 2 + 2x_1 + x_5 + x_4
\end{aligned}
$$

This is the optimal solution since all the basic variables are positive.

7.4 PARAMETRIC PROGRAMMING

We may be interested in mapping out how the solution to a linear programming problem changes as the problem data change in some systematic way. This is known as *parametric programming* and there are two basic forms: *parametric variation of the objective function* and *parametric variation of the right-hand side*. The former is closely related to the primal (or original) simplex method and the latter to the dual simplex method in the same way.

The algorithm for a *parametric objective function variation* can be derived as follows. We wish to maximize $x_0 + \theta x_0^*$ for all values of the parameter θ between 0 and θ_{MAX}. We may assume that $0 < \theta_{\text{MAX}} \le \infty$. We begin by solving the problem for $\theta = 0$ and consider the tableau

$$x_0 = \bar{a}_{00} - \sum_{j=1}^{n-m} \bar{a}_{0j} X_{m+j}$$

$$x_0^* = \bar{a}_{00}^* - \sum_{j=1}^{n-m} \bar{a}_{0j}^* X_{m+j}$$

$$X_i = \bar{a}_{i0} - \sum_{j=1}^{n-m} \bar{a}_{ij} X_{m+j} \qquad (i = 1, \dots, m)$$

We see that

$$x_0 + \theta x_0^* = (\bar{a}_{00} + \theta \bar{a}_{00}^*) - \sum_{j=1}^{n-m} (\bar{a}_{0j} + \theta \bar{a}_{0j}^*) X_{m+j}$$

so the solution for $\theta = 0$ remains optimum as θ increases, provided that

$$\bar{a}_{0j} + \theta \bar{a}_{0j}^* \geqslant 0 \qquad (j = 1, \dots, n - m) \tag{7.4.1}$$

If $\bar{a}_{0j}^* \geqslant 0$ for all j this will always be true. However, if one of the $\bar{a}_{0j}^*$ is negative, say $\bar{a}_{0k}^*$, there will be some positive value of θ which causes expression (7.4.1) for $j = k$ to become negative. So we compute the ratios

$$\frac{\bar{a}_{0j}}{|\bar{a}_{0j}^*|} \qquad \text{(for all } j \text{ such that } \bar{a}_{0j}^* < 0)$$

and note the value of j, say q, which has the minimum ratio value. Therefore setting

$$\theta_{\mathrm{C}} = \frac{\bar{a}_{0q}}{|\bar{a}_{0q}^*|}$$

we know that when $\theta = \theta_{\mathrm{C}}$ the solution is no longer optimal and we must make a pivot step. In other words, we make X_{m+q} basic and choose the variable to be removed from the basis in the usual way for the simplex method, i.e. X_p, where p is the argument which has the minimum ratio value

$$\frac{\bar{a}_{i0}}{\bar{a}_{iq}} \qquad \text{(for all } i \text{ where } \bar{a}_{iq} > 0)$$

After pivoting we have a new tableau, which is optimum for the next range of values of θ. The process continues until one of three things happens:

(1) $\theta_{\mathrm{C}} \geqslant \theta_{\mathrm{MAX}}$, in which case the current trial solution remains optimum until the end of the prescribed range of values of θ.
(2) $\bar{a}_{0j}^* \geqslant 0$ for all j. This means that the current trial solution remains optimum for all greater values of θ.
(3) $\bar{a}_{iq} \leqslant 0$ for all i. This means that for any $\theta > \theta_{\mathrm{C}}$ the objective function is unbounded.

The algorithm for *parametric right-hand side variation* can be derived as follows. Again we wish to maximize x_0 for all values of the parameter θ between 0 and θ_{MAX}, where $0 < \theta_{MAX} \leqslant \infty$. This time we consider the tableau

$$x_0 = \bar{a}_{00} + \theta \bar{a}_{00}^* - \sum_{j=1}^{n-m} \bar{a}_{0j} X_{m+j}$$

$$X_i = \bar{a}_{i0} + \theta \bar{a}_{i0}^* - \sum_{j=1}^{n-m} \bar{a}_{ij} X_{m+j} \qquad (i = 1, \ldots, m)$$

We first solve the problem for $\theta = 0$ and we see that, if $\bar{a}_{0j} \geqslant 0$ for all j, the solution given by

$$X_i = \bar{a}_{i0} + \theta \bar{a}_{i0}^* \qquad (i = 1, \ldots, m)$$

is optimum provided that all the above X_i are non-negative. Thus we compute the minimum of the ratio values

$$\theta_C = \min_i \frac{\bar{a}_{i0}}{|\bar{a}_{i0}^*|} \qquad (\text{for all } i \text{ such that } \bar{a}_{i0}^* < 0)$$

and note the argument i, say p, which attains this minimum value. Therefore we know that when $\theta = \theta_C$ we must make a pivot step, making X_p non-basic, and choosing the variable to enter the basis in the usual way for the dual simplex method, i.e. X_{m+q}, where q is the argument with the minimum value of the ratios

$$\frac{\bar{a}_{0j}}{|\bar{a}_{pj}|} \qquad (\text{for all } j \text{ such that } \bar{a}_{pj} < 0)$$

After pivoting we have a new tableau, which is optimum for the next range of values of θ. The process continues until one of three things happens:

(1) $\theta_C \geqslant \theta_{MAX}$, in which case the current basis remains optimum until the end of the prescribed range of values of θ.
(2) $\bar{a}_{i0}^* \geqslant 0$ for all i. This means that the current basis remains optimum for all greater values of θ.
(3) $\bar{a}_{pj} \geqslant 0$ for all j. This means that for any $\theta > \theta_C$ the problem is infeasible.

8

How to Apply Linear Programming

8.1 INTRODUCTION: MATHEMATICAL PROGRAMMING SYSTEMS

So far we have studied small examples of linear programming problems which we were able to solve manually using the simplex and dual simplex methods. In practice, however, linear programming problems are usually much larger than our examples, often having thousands of constraints and variables. To solve these large problems efficient computer programs have been developed and are now widely available for many computers. They are generally known as *mathematical programming systems.*

Computer programs to solve linear programming problems by the simplex method have existed since the early 1950s. Initially they were only capable of solving small problems, but since then there has been a steady increase in the size of problem that can be solved. This has been due as much to a better understanding of how to exploit sparseness as to larger and faster computers. However, as Dantzig (1963) makes clear, the development of computers was crucial to the development of mathematical programming, and the two have evolved in parallel over the last 30 years.

The permitted number of constraints (other than simple bounds on individual variables) reached a thousand in the early 1960s, and there was no definite limit on the number of variables. In principle, current algorithms impose no definite limit on the numbers of either constraints or variables. In practice, problems with over two thousand constraints are still considered large and may prove expensive in computer time to solve. In some systems the number of constraints plus variables is limited to 32 000 (the largest signed integer that can be stored in 2 bytes of computer memory).

Orchard Hays (1978a–c) writes about the first decade of computational linear programming software (or 'programs') with unique authority. The fact that existing systems are often descendants of others is of some importance, and, as with other computer programs, modifications become progressively more difficult as the concepts develop further from those used when the system was first designed. Examples of mathematical programming systems are MPS and its powerful successors MPSX and MPSX/370 for the IBM 360 and 370 series and their successors, and SCICONIC, which is available on a wide range of machines, from very powerful mainframes to the latest personal computers (PCs). These systems are all capable of solving large linear programming problems provided they are reasonably sparse. In other words, the number of non-zero elements in the mathematical formulation of the problem is now more significant than the numbers of constraints and variables.

Besides being able to solve linear programming problems, most systems also contain algorithms for handling integer variables and some non-linear problems. Hence the name 'mathematical programming systems' instead of just 'linear programming systems'.

How we communicate the mathematical programming problem to the computer is obviously of some importance, and is described in Section 8.4. First, we will review how and why linear programming is used and develop a systematic approach to the formulation of mathematical programming models.

8.2 APPLICATIONS OF MATHEMATICAL PROGRAMMING

So far we have concentrated on linear programming. An extension of this is the more general mathematical programming:

minimize $\qquad f(\mathbf{x})$
subject to

$$g_i(\mathbf{x}) \leqslant b_i \qquad (i = 1, \ldots, m)$$
$$x_j \geqslant 0 \qquad (j = 1, \ldots, n)$$

where the functions $f(\mathbf{x})$ and $g_i(\mathbf{x})$ may be non-linear, or where the variables x_j may be required to take integer values. We will look at non-linear programming and integer programming in greater detail in Chapters 10 and 11, but for now it is sufficient that we realize that the following remarks about modelling and applications apply to mathematical programming in general.

In general terms mathematical programming is concerned with the best way to allocate scarce resources to alternative activities. For example, the single-mix blending problem (Section 5.2) was concerned with producing the cheapest animal food mixture possible from limited supplies of raw materials such as sago flour or meat meal. Williams (1978) lists applications of mathematical programming (and refers to publications) under the following headings:

Petroleum industry
Chemical industry
Manufacturing
Transport
Finance
Agriculture
Health
Mining

Manpower planning
Food
Energy
Pulp and paper
Advertising
Defence
Other applications

Although many applications have special features, the main elements of the model are usually combinations of the concepts which will be introduced in the examples in Chapter 9. Once these concepts have been mastered, we will be some way towards understanding both how and why mathematical programming can be usefully applied in other contexts. It is important, however, to realize why mathematical programming applications have been successful. First, mathematical programming gives true optimum solutions to problems to which we could otherwise only find approximate solutions. Second, the concepts of mathematical programming—the quantification of the objectives and the set of all possible ways of achieving them—provide a framework for thinking about all the relevant data, and an occasion for collecting them. However, the consequences of these data is also important. Often the only way to achieve a realistic set of data is to show the people who collected them the consequences of their initial data, after the model has been solved.

It is useful to distinguish between established and new mathematical programming models. An *established* model is run from time to time with updated data as part of some operational decision-making routine. The purpose is then to suggest a specific course of action to management, and the suggestion will usually be accepted. A *new* model may also be used in this way but is more often employed to gain greater understanding of the situation. It may be run under a variety of assumptions that lead to different conclusions, and the model itself will not suggest which set of assumptions is most appropriate.

During the model development and data-gathering phase we must therefore be prepared to make many optimization calculations which can be shown to management to see if they are sensible. If what the model recommends is not considered sensible, we have to find out why it is not acceptable. Neither the analyst or the manager should accept the recommendations from the model unless they can be explained qualitatively as the natural consequences of the physical and economic assumptions. We can paraphrase this by saying that the results should only be accepted *if they are obvious*. The reader may think that the model is then of no real use! This, however, is not so, because many things are obvious once someone has pointed them out, when they were not at all obvious beforehand.

8.3 A TECHNIQUE FOR MODEL DOCUMENTATION

The task of defining a linear programming problem can be illuminated by considering an example that occurs as part or all of many real applications: this is the so-called *transportation problem*. Specified quantities of some homogeneous material are available at a number of sources, and specified quantities must be delivered to each of a number of destinations. The cost of supplying each destination from each source is defined in £/tonne. What is the cheapest way to meet all the requirements?

The problem is clearly to minimize a function of variables representing the number of tonnes sent from each source to each destination. Our first task is to choose what to call these variables. It is not convenient to use a different letter for each variable. This is partly because there may well not be enough letters but, more fundamentally, because we need a notation that indicates the relationships between different variables, so we use subscripts. This problem can be naturally formulated in terms of just two subscripts, say,

i for sources

and

j for destinations

We can now write x_{ij} for the variable defining the number of tonnes sent from source i to destination j. So x_{11} denotes the number of tonnes sent from source 1 to destination 1, and x_{23} denotes the number of tonnes sent from source 2 to destination 3.

The data can also be defined in terms of these subscripts. Let I denote the number of sources and J the number of destinations. Let A_i denote the number of tonnes available at source i, D_j the number of tonnes required at destination j and C_{ij} the cost in £/tonne of supplying destination j from source i.

Suppose that $I = 2$ and $J = 3$. Then we can define the problem as follows:

minimize

$$C_{11}x_{11} + C_{12}x_{12} + C_{13}x_{13} + C_{21}x_{21} + C_{22}x_{22} + C_{23}x_{23}$$

subject to the constraints that

(1) The amounts supplied from each source do not exceed the amounts available, i.e.

$$x_{11} + x_{12} + x_{13} \leqslant A_1$$
$$x_{21} + x_{22} + x_{23} \leqslant A_2$$

(2) The requirements at each destination are met, i.e.

$$x_{11} + x_{21} = D_1$$
$$x_{12} + x_{22} = D_2$$
$$x_{13} + x_{23} = D_3$$

(3) All variables are non-negative, i.e. greater than or equal to zero. Note that, even though we might interpret $x_{11} = -1$ as sending one tonne back from destination 1 to source 1, this would not produce a revenue of £C_{11}.

Note that we can write the formulation more compactly by exploiting our subscript notation more fully. The two availability constraints are really the same except for the value of i. We can thus write them as

$$x_{i1} + x_{i2} + x_{i3} \leqslant A_i \qquad \text{(for all } i)$$

Similarly, the demand constraints can be written as

$$x_{1j} + x_{2j} = D_j \qquad \text{(for all } j)$$

The notation becomes even more compact if we use summation signs to represent the additions on the left-hand side of these constraints. Specifically, we can write

$$\sum_j x_{ij} \leqslant A_i \qquad \text{(for all } i)$$

and

$$\sum_i x_{ij} = D_j \qquad \text{(for all } j)$$

The objective function can be written as

$$\sum_j C_{1j} x_{1j} + \sum_j C_{2j} x_{2j}$$

and this in turn can be further simplified by summing over i. Therefore we have to minimize

$$\sum_i \sum_j C_{ij} x_{ij}$$

subject to

$$\sum_j x_{ij} \leqslant A_i \qquad \text{(for all } i)$$

$$\sum_i x_{ij} = D_j \qquad \text{(for all } j)$$

$$x_{ij} \geqslant 0 \qquad \text{(for all } i, j)$$

These four lines of algebra define a transportation problem for any values of I and J. The problem might have 10 sources and 200 destinations, when it has 2000 variables and 210 constraints, excluding the constraints that $x_{ij} \geqslant 0$ for all i, j, which are usually taken for granted.

This example illustrates two important points:

(1) Practical linear programming formulations can all too easily require hundreds of constraints and thousands of variables; while
(2) The algebraic formulation is precise and often compact.

It is therefore natural to base the computer implementation of a mathematical programming model on an algebraic formulation. However, the model structure is rarely as simple as in the above example, and the algebraic formulation itself can easily become unintelligible if it is not set out systematically. Beale *et al.* (1974) therefore recommend the following order of presentation:

(1) *Subscripts.* Use lower-case letters, with the corresponding capital letter representing the maximum possible value. These define the classes of objects in the model, e.g. sources, *i*, and destinations, *j*, in the above example.
(2) *Sets.* These may be needed to give precise definitions of ranges of summation of a subscript, or conditions under which a constraint, variable or constant is defined. Use capital letters.
(3) *Constants.* These are the known data when the model is to be optimized. Denote them by using capital letters.
(4) *Variables.* These are the linear programming variables. Use lower-case letters.
(5) *Constraints (including the objective function).* Use algebraic notation.

Never use one letter to mean two different things, but feel free to develop composite letters by using capital letters as literal subscripts that never take numerical values. On the other hand, lower-case subscripts always take numerical values: for example,

A_{Vrs} might represent the amount of material *r* available from source *s* (in tonnes), while

x_{Ah} might represent the rate of consumption of feed material *h* (in tonnes/day).

This approach may seem less straightforward than starting with the constraints and then explaining the symbols that occur in them, but it is recommended for the following reasons.

A definition of all subscripts defines the scope of the model. If the subscripts are used consistently, the definition of subscripted constants and variables become clearer and can be more compact, as illustrated earlier.

Sets may not be needed on simple models, but they often provide the only compact way to record a precise definition of the domain of a summation, or the conditions for the existence of a variable or constraint. For example, in the transportation problem some sources may not be able to supply all destinations, so we may define S_{Ji} as the set of destinations that can be supplied from source *i*, and S_{1j} as the set of sources that can supply destination *j*. The constraints then become

$$\sum_{j \in S_{Ji}} x_{ij} \leqslant A_i \qquad \text{(for all } i\text{)}$$

and

$$\sum_{i \in S_{1j}} x_{ij} = D_j \qquad \text{(for all } j\text{)}$$

The constants are part of the definition of the problem, so they logically precede the decision variables, which represent an approach to its solution.

The notation should distinguish clearly between the constants, or assumed quantities, and the decision variables that are to be determined by the model. This distinction may not be clear from the context. For example, the amounts of each product to be made may be either data or variables. Possible confusion is avoided by always using capital letters for constants and lower-case ones for decision variables.

8.4 COMMUNICATING THE FORMULATION TO THE COMPUTER

Once the mathematical programming model has been formulated we must communicate it to the computer program in a form it can understand. Nearly all mathematical programming systems use a standard input format, known as *MPS format*, which was developed by IBM for their MPS 360 system.

First the constraints or *rows* of the problem are listed, followed by the non-zero coefficients in these constraints, sorted by variable, or *column*. Finally, for each of the constraints, the non-zero right-hand side values are listed.

For example, to solve the following small transportation problem, with two sources and three destinations, on a computer we would need to generate the file or *matrix* shown in Figure 8.4.1:

minimize the objective function:

$$x_{11} + 3x_{12} + 4x_{13} + 2x_{21} + 5x_{22} + 7x_{23} \quad \text{(OBJ)}$$

subject to the availability constraints at both sources:

$$
\begin{aligned}
x_{11} + x_{12} + x_{13} & \leqslant 5.0 \quad \text{(AV1)} \\
x_{21} + x_{22} + x_{23} & \leqslant 1.0 \quad \text{(AV2)}
\end{aligned}
$$

and subject to the demand constraints at the three destinations:

$$
\begin{aligned}
x_{11} \quad\quad\quad + x_{21} \quad\quad\quad & = 3.0 \quad \text{(DE1)} \\
x_{12} \quad\quad\quad + x_{22} \quad & = 3.0 \quad \text{(DE2)} \\
x_{13} \quad\quad\quad + x_{23} & = 3.0 \quad \text{(DE3)}
\end{aligned}
$$

There is clearly a large element of repetition in the preparation of such data, which becomes even more painful with larger problems. Given the mathematical formulation it is a fairly straightforward task to write a computer program to read in the constants (for example, the demands of 3.0 in the transportation problem) and produce the required input file or *matrix* for the mathematical programming system. Such a program is called a *matrix generator*. Indeed the task is so straightforward that it can be further automated. For example, Scicon has developed a *matrix generator generator* program, known as MGG, which takes a mathematical formulation very similar to the algebraic formulation considered in Section 8.3 and produces the actual matrix generator program. The numerical data for a particular problem are then fed to the matrix generator program to produce the required matrix. Also the same system can be used to generate a *report writer* program to analyse the solution produced by the mathematical programming system. Consequently, MGG or a similar *modelling*

```
              NAME            EXAMPLE
              ROWS
               N   OBJ
               L   AV1
               L   AV2
               E   DE1
               E   DE2
               E   DE3
              COLUMNS
                     X11      OBJ       1.0
                     X11      AV1       1.0
                     X11      DE1       1.0
                     X12      OBJ       3.0
                     X12      AV1       1.0
                     X12      DE2       1.0
                     X13      OBJ       4.0
                     X13      AV1       1.0
                     X13      DE3       1.0
                     X21      OBJ       2.0
                     X21      AV2       1.0
                     X21      DE1       1.0
                     X22      OBJ       5.0
                     X22      AV2       1.0
                     X22      DE2       1.0
                     X23      OBJ       7.0
                     X23      AV2       1.0
                     X23      DE3       1.0
              RHS
                              AV1       5.0
                              AV2       1.0
                              DE1       3.0
                              DE2       3.0
                              DE3       3.0
              ENDATA
```

Figure 8.4.1

language allows the analyst to specify his problem in algebraic form and
automatically generates the programs he needs to solve the problem and report
the solution. Fourer (1983) reviews the different matrix generators and modelling
languages available.

9

Examples of Linear Programming Problems

9.1 PREFACE TO EXAMPLES

In the following examples the essence of the problem to be solved is defined by the subscripts and constants. The definitions of the constants can be very compact when one relies on the subscripts being used consistently, but it is important to define the units in which all constants and variables are measured. This avoids confusion in the user's mind about the correct numbers to define some quantity, and can also help to do the same for the analyst's mind when formulating the problem, since it helps him to check that the formulation is dimensionally correct. For example, the user obviously needs to know whether a price is measured in £/tonne or £/m^3 or \$/tonne or \$/m^3. Also, if some quality blends by volume while the quantities of different materials in the blend are measured in tonnes, the explicit statement of this fact helps to remind the analyst to divide the quantities by a density to convert them to volumes before multiplying by the appropriate quality indices.

9.2 A MULTI-MIX BLENDING PROBLEM

In Section 5.2 we studied the single-mix blending problem, where a single product has to be made from a given pool of raw materials, but, in practice, blenders have to make several products from the same pool of raw materials. These multi-product problems can sometimes be treated separately, considering just one product at a time, as we did in the single-mix blending problem. However, the problem must be considered as a whole if there is a limit on the availabilities of some materials that have to be shared between different products. A further complication arises when some materials are available in different quantities at different prices. In particular, the manager may assign a lower price to the use of existing stocks of materials than to the use of material that has to be newly purchased. The formulation can be described in the following ways.

Subscripts

p	Product
q	Quality
r	Material
s	Source

Sets

S_{Rp}	Set of materials that can be used to make product p
S_{Pr}	Set of products in which material r can be used

Constants

A_{qr}	Amount of q per tonne of r (fraction)
A_{Vrs}	Availability of material r from source s (tonnes)
D_p	Demand for product (tonnes)
P_{rs}	Price of raw material (£/tonne)
Q_{MAXpq}	Maximum permitted amount of q in product p (fraction)
Q_{MINpq}	Minimum permitted amount of q in product p (fraction)

The natural formulation for this problem is to write x_{rs} for the amount of raw material purchased (tonnes) and y_{pr} for the amount of raw material used (tonnes). Then we must minimize

$$\sum_r \sum_s P_{rs} x_{rs}$$

subject to quality constraints that

$$\sum_{r \in S_{Rp}} (A_{qr} - Q_{MAXpq}) y_{pr} \leqslant 0 \qquad \text{(for all } p, q)$$

$$\sum_{r \in S_{Rp}} (A_{qr} - Q_{MINpq}) y_{pr} \geqslant 0 \qquad \text{(for all } p, q)$$

the availability constraints that

$$x_{rs} \leqslant A_{Vrs} \qquad \text{(for all } r, s)$$

and the material balance constraints that

$$\sum_s x_{rs} - \sum_{p \in S_{Pr}} y_{pr} = 0 \qquad \text{(for all } r)$$

$$\sum_{r \in S_{Rp}} y_{rp} = D_p \qquad \text{(for all } p)$$

together with the constraints that all variables are non-negative.

It is of some interest to note that this problem has an apparently simpler formulation in terms of a single type of variable. If z_{prs} is the number of tonnes of material r bought from source s and used to make product p, the problem is to minimize

$$\sum_p \sum_r \sum_s P_{rs} z_{prs}$$

subject to the constraints

$$\sum_{r \in S_{Rp}} \sum_s (A_{rq} - Q_{\text{MAX}pq}) z_{prs} \leqslant 0 \qquad \text{(for all } p, q)$$

$$\sum_{r \in S_{Rp}} \sum_s (A_{rq} - Q_{\text{MIN}pq}) z_{prs} \geqslant 0 \qquad \text{(for all } p, q)$$

$$\sum_p z_{prs} \qquad\qquad\qquad \leqslant A_{\text{V}rs} \qquad \text{(for all } r, s)$$

$$\sum_{r \in S_{Rp}} \sum_s z_{prs} \qquad\qquad = D_p \qquad \text{(for all } p)$$

This alternative formulation omits one set of material balance constraints, but it has more decision variables, and, more seriously, more non-zero coefficients than the recommended formulation. These are some of the practical considerations that arise when developing mathematical programming models.

9.3 A PROCESSING PROBLEM

Many industrial plants contain a number of processing units, each of which can be operated in one or more modes. Each mode uses one or more types of raw material and produces one or more products. To illustrate the types of constraint arising in such models we consider a simple problem defined by the following data.

Subscripts

h	Material
i	Unit
m	Mode of operation

Sets

S_{HB}	Set of materials that can be bought
S_{HS}	Set of materials that can be sold

Constants

A_{him} Net production of material h when unit i is operated in mode m (tonnes/day). This is negative if the process uses material h as an input and positive if the process produces material h as an output.

D_{AYS} Duration of time period (days)
D_{MAXh} Maximum demand for material h ($h \in S_{HS}$) (tonnes)
D_{MINh} Minimum demand for material h ($h \in S_{HS}$) (tonnes)
P_{Bh} Buying price for material h ($h \in S_{HB}$) (£/tonne)
P_{Sh} Selling price for material h ($h \in S_{HS}$) (£/tonne)

The problem is to maximize the revenue from sales less the cost of bought in materials without exceeding the productive capacity of any unit. The natural formulation is to use three sets of variables:

b_h Number of tonnes of material h bought ($h \in S_{HB}$) (tonnes)
s_h Number of tonnes of material h sold ($h \in S_{HS}$) (tonnes)
x_{im} Number of days during which unit i operates in mode m.

Then the problem is to maximize

$$\sum_{h \in S_{HS}} P_{Sh}s_h - \sum_{j \in S_{HB}} P_{Bh}b_h$$

subject to the capacity constraints

$$\sum_m x_{im} \leqslant D_{AYS} \qquad \text{(for all } i\text{)}$$

and the material balance constraints

$$b_h + \sum_i \sum_m A_{him}x_{im} - s_h = 0 \qquad \text{(for all } h\text{)}$$

(where the term b_h exists only if $h \in S_{HB}$ and s_h exists only if $h \in S_{HS}$) and the demand constraints

$$D_{MINh} \leqslant s_h \leqslant D_{MAXh} \qquad \text{(for all } h \in S_{HS}\text{)}$$

There are a number of points to note about this formulation. By using the general notion of material the formulation is clearly more concise than if input material, intermediate material and finished product had been used specifically. It can be seen that no data have been provided for costs or capacities of transporting material within the plant, so these activities are assumed to be effectively unlimited and to have a negligible cost. Similarly, the operating costs of the units are not specified, so they are assumed to be independent of how the units are used.

If we want to limit the amount of material that can be transmitted from some unit, say X, to some other unit, say Y, then the simplest approach is to define the pipework connecting these units as another unit, say P, with a specified capacity. It then becomes necessary to define the output material from P as mathematically different from the input material, even though they are physically the same. The unit P then converts input material into output material, and Y can only use output material from P.

Note also that there is a slack on the capacity constraints, indicating that the units do not have to be used all the time, but there is no slack on the material balance constraints. This is important, because some by-products may be surplus to requirements and may have to be processed to turn them into something else that can be sold. Some products may have a negative selling price, indicating that we have to pay someone to dispose of them. Note particularly that the processing variables are x_{im} and not x_{him}: the amounts of material either used or produced are proportional to the overall level of each production activity and are not independent variables. This is different from blending problems, although one can think of blending problems in this way, with each mode making the product from a single material and with quality constraints restricting the choice of combinations of modes. In practice, a model may contain processing activities of the type considered in this example and also blending activities.

9.4   A MULTI-TIME-PERIOD PRODUCTION AND STORAGE PROBLEM

Consider a problem defined by the following data.

Subscripts

m	Machine
p	Product
r	Raw material
t	Time period

Sets

S_{Mp}	Set of machines on which product p can be made
S_{Pm}	Set of products that can be made on machine m

Constants

C_{rt}	Expected cost of raw material (£/tonne)
C_{HPp}	Cost of holding 1 tonne of product from one time period until the next (£)
C_{HRr}	Cost of holding 1 tonne of raw material from one time period until the next (£)
D_{pt}	Demand (tonnes)
H	Number of working hours per time period
Q_{rp}	Number of tonnes of raw material r used in the manufacture of 1 tonne of p (fixed formula)
R_{mp}	Rate of production of p (tonnes/hour)
S_{POp}	Initial stock of product (tonnes)
S_{ROr}	Initial stock of raw material (tonnes)

The problem is to minimize the cost of purchasing raw materials plus the costs of holding raw materials and products while meeting the demands for all products in all time periods. The natural formulation is then to use four sets of variables as follows:

x_{Rrt} Amount of raw material purchased (tonnes)

S_{Rtr} Amount of raw material in store at *end* of time period t (tonnes)

s_{Ptp} Amount of product p in store at *end* of time period t (tonnes)

z_{mpt} Amount of product p made on machine m during time period t (tonnes). Note that, since the formula is fixed, there is a single production variable for each machine and we do *not* need separate variables for each raw material used to make the product.

Then the problem is to minimize

$$\sum_r \sum_t C_{rt} x_{Rrt} + \sum_r \sum_t C_{HRr} s_{Rtr}$$

$$+ \sum_p \sum_t C_{HPp} s_{Ptp}$$

subject to

$$\sum_{p \in S_{Pm}} \frac{1}{R_{mp}} z_{mpt} \leqslant H \qquad \text{(for all } m, t)$$

$$s_{Rt-1,r} + x_{Rrt} - \sum_m \sum_p Q_{rp} z_{mpt} - s_{Rtr} = 0 \qquad \text{(for all } r, t)$$

$$s_{Pt-1,p} + \sum_{m \in S_{Mp}} z_{mpt} - s_{Ptp} = D_{pt} \qquad \text{(for all } p, t).$$

The last two equations are different for $t = 1$ in that the first term is now a constant and appears on the other side.

9.5 A LONG-TERM INVESTMENT PLANNING PROBLEM

Consider a problem of long-term investment planning for electricity supply.

Subscripts

l Demand level ($= 1$ for lowest level in year
                              $= L$ for highest level in year)

s Source (i.e. generator)

t Time period (year)

Constants

C_{CAPs} Capital cost per unit of additional capacity (incurred in year before commissioning) (£/W)

C_{Fs} Fixed annual cost for maintaining a unit of capacity (£/W)

C_{Vs} Variable cost of using generator (£/W year)

D_{lt} Demand (MW)

F_l Fraction of year for which demand is at level l

Q_{0s} Initial capacity of source (MW)

α Discount factor

The problem is to minimize the total cost of meeting the demands at all times (i.e. at all levels) in all future years, where future costs are discounted at a rate α, so the effective cost of £1 spent in year t is

$$\text{£}\frac{1}{(1+\alpha)^t}$$

The natural formulation is as follows:

Variables

x_{lst} Amount of capacity used from source s when the demand is at level l in time period t (MW)

y_{st} Amount of new capacity installed (MW)

Constraints

Minimize

$$\sum_s \sum_t \left(\sum_{t_1=t}^{T} \frac{1}{(1+\alpha)^{t_1}} \right) (\alpha C_{\text{CAP}s} + C_{\text{F}s}) y_{st}$$

$$+ \sum_l \sum_s \sum_t \frac{1}{(1+\alpha)^t} F_l C_{\text{V}s} x_{lst}$$

subject to

$$x_{lst} - \sum_{t_1=1}^{t} y_{st_1} \leqslant Q_{0s} \qquad \text{(for all } l, s, t)$$

$$\sum_s x_{lst} = D_{lt} \qquad \text{(for all } l, t)$$

In this formulation the capital costs are not paid as such. Instead the money is considered to be borrowed, and interest is paid annually, at a rate α. This implicitly values the capital assets at the end of the time horizon at cost, i.e. without any depreciation. An alternative is to value it at zero, in which case the coefficient of y_{st} becomes

$$\frac{1}{(1+\alpha)^{t-1}} C_{\text{AP}s} + \sum_{t_1=t}^{T} \frac{1}{(1+\alpha)^{t_1}} C_{\text{F}s}$$

In practice these 'end effects' should not be important if T is large enough. On the other hand, in some contexts one must consider the tax implications of different types of expenditure, and these may greatly affect the relative effects of different types of expenditure on the organization.

The objection to this formulation is that there may be an excessive number of these capacity constraints. We can make the formulation more compact by relying on the fact that, in practice,

$$x_{1st} \leqslant x_{2st} \leqslant \cdots \leqslant x_{Lst}$$

Thus we write

$$z_{st} = x_{lst} - x_{l-1,st}$$

We can now eliminate the variables x_{st} from the formulation and write:

minimize

$$\sum_s \sum_t \left(\sum_{t_1=t}^{T} \frac{1}{(1+\alpha)^{t_1}} \right) (\alpha C_{\text{CAPs}} + C_{\text{Fs}}) y_{st}$$

$$+ \sum_l \sum_s \sum_t \frac{1}{(1+\alpha)^t} \left(\sum_{l_1=l}^{L} F_{l_1} \right) C_{\text{VS}} z_{lst}$$

subject to

$$\sum_{l=1}^{L} z_{lst} - \sum_{t_1=1}^{t} y_{st_1} \leqslant Q_{0s} \qquad \text{(for all } s, t)$$

$$\sum_{l_1=1}^{l} \sum_s z_{l_1 st} = D_{lt} \qquad \text{(for all } l, t)$$

This last set of constraints is now unnecessarily dense, and this can be seen from a 'picture' of the matrix which contains triangular arrays of 1's. This density can be reduced by subtracting each row from the following one, and the result of this subtraction can be interpreted as saying that each constraint should represent the incremental change from one condition to the next, and not the cumulative change from the original conditions.

This means that the last set of constraints reads

$$\sum_s z_{lst} = D_{lt} - D_{l-1,t} \qquad \text{(for all } l, t)$$

A similar approach is often appropriate for capacity constraints, but here it would make the model more dense unless L is small in comparison with T.

PART 3

CONSTRAINED OPTIMIZATION: NON-LINEAR AND DISCRETE

10
Non-Linear Programming

10.1 INTRODUCTION

This chapter is concerned with non-linear programming: that is, with essentially linear programming problems with some non-linear constraints, or a non-linear objective function, or both.

We will begin by reviewing the basic theory of optimization from a more general point of view than was convenient when introducing linear programming. In particular, we will look at *Lagrange multipliers* and the famous *Kuhn–Tucker conditions* which are necessary for a local minimum. This will give us a very basic method for solving non-linear programming problems, although in practice very few problems can be solved using these Lagrangian equations.

We will then go on to study rather more useful methods for solving non-linear programming problems. Many methods have been devised to solve problems of this type (see Powell, 1982), but their success depends very much on the structure of the non-linear problem and the degree of the non-linearity. Instead of describing all the approaches which have been adopted in the past (many of which are no longer used) we will concentrate on two methods which are in practical use. The first is based on the concepts of *separable programming* and the other on the *reduced gradient method*.

10.2 LAGRANGE MULTIPLIERS AND KUHN–TUCKER CONDITIONS

Suppose initially that we just wish to minimize an unconstrained differentiable function $f(x)$ which is defined for all values of the single variable x. Then it is well known that a necessary condition for a minimum is that $df/dx = 0$. A value of x for which $df/dx = 0$ does not necessarily minimize $f(x)$ (it may maximize it or else be a saddle point), but under certain conditions it will definitely minimize the function. One such condition arises if the function $f(x)$ is *convex* (see Section 1.2 for the definition of a convex function).

A precisely analogous situation arises if we wish to minimize an unconstrained differentiable function of several variables, say $f(x_1, \ldots, x_n)$. A necessary condition for a minimum now is that

$$\frac{\partial f}{\partial x_1} = \frac{\partial f}{\partial x_2} = , \ldots, \frac{\partial f}{\partial x_n} = 0$$

Again, if $f(\mathbf{x})$ is a convex function, any point where all these partial derivatives vanish must be a minimum.

The problem becomes a little more difficult if we have to consider equality constraints. Suppose that we want to minimize the function $f(x_1, \ldots, x_n)$ subject to the equality constraints

$$g_i(x_1, \ldots, x_n) = b_i \qquad (i = 1, \ldots, m) \tag{10.2.1}$$

We may be able to overcome the problem by using the constraints to solve for m of the variables in terms of the others. If this can be done it reduces the dimension of the problem and the constraints have actually made things easier, but very often this is not the case. For example, with the constraint

$$\sum_{j=1}^{n} x_j^2 = 1$$

we can write

$$x_n = \pm \left(1 - \sum_{j=1}^{n-1} x_j^2\right)^{1/2}$$

However, this is not only awkward because of the multiple solutions but also obscures the implicit constraint on the independent variables $x_1, \ldots, x_{n-1}$ to the effect that

$$\sum_{j=1}^{n-1} x_j^2 \leqslant 1$$

So what else can we try in the case of equality constraints? One approach is to solve the problem using Lagrange multipliers as follows.

Since each function $g_i(\mathbf{x})$ must equal b_i the problem is unaffected if we replace the objective function $f(\mathbf{x})$ to be minimized by the *Lagrangian function*

$$F(\mathbf{x}) = f(\mathbf{x}) + \sum_{i=1}^{m} \pi_i(g_i(\mathbf{x}) - b_i) \tag{10.2.2}$$

for any values of the quantities $\pi_1, \ldots, \pi_m$, which are called *Lagrange multipliers*. The Lagrangian function $F(\mathbf{x})$ is often written as $L(\mathbf{x}, \pi)$ to emphasize its dependence on both $\mathbf{x}$ and π.

The useful fact about Lagrange multipliers is that if we can find values of $\pi_1, \ldots, \pi_m$ such that the point $\mathbf{x}$ where $F(\mathbf{x})$ is minimized happens to satisfy all the constraints $g_i(\mathbf{x}) = b_i$, then we will also have solved our original constrained problem. The proof is obvious, since the Lagrangian function (10.2.2) reduces to our original objective function $f(\mathbf{x})$ if the constraints are satisfied.

Thus we try to find points $\mathbf{x}, \pi_1, \ldots, \pi_m$ which minimize equation (10.2.2) and satisfy constraints (10.2.1). From the earlier discussion about minimization we know that a necessary condition for a minimum is that $\partial F / \partial x_j = 0$ $(j = 1, \ldots, n)$.

In other words, we require that

$$\frac{\partial f}{\partial x_j} + \sum_{i=1}^{m} \pi_i \frac{\partial g_i}{\partial x_j} = 0 \qquad (j=1,\ldots,n)$$

$$g_i(x_1,\ldots,x_n) = b_i \qquad (i=1,\ldots,m)$$

So we now have $n+m$ equations for the $n+m$ unknowns $x_1,\ldots,x_n$ and $\pi_1,\ldots,\pi_m$ which we may possibly be able to solve.

Lagrange multipliers can be generalized to apply to inequality constraints $g_i(\mathbf{x}) \leqslant b_i$. To see intuitively how this works, it is perhaps easier to think again of functions of a single variable. We know that the minimum of some function $f(x)$, subject to the constraint $x \leqslant b$, will be attained either when $df/dx = 0$ or when $df/dx < 0$ and $x = b$. This can be expressed by the conditions $df/dx + \pi = 0$, $\pi \geqslant 0$ and $\pi(x - b) = 0$. The corresponding condition on the Lagrange multiplier π_i when the constraint $g_i(\mathbf{x}) = b_i$ is replaced by $g_i(\mathbf{x}) \leqslant b_i$ is that either $\pi_i = 0$ or $\pi_i > 0$ and $g_i(\mathbf{x}) = b_i$, i.e. that $\pi_i(g_i(\mathbf{x}) - b_i) = 0$.

If all the constraints are of the form $g_i(\mathbf{x}) \leqslant b_i$, and the functions $g_i(\mathbf{x})$ are all convex and differentiable, we obtain the famous *Kuhn–Tucker conditions* (Kuhn and Tucker, 1951) to the effect that $\mathbf{x}$ minimizes $f(\mathbf{x})$ subject to the constraints if and only if there exist values of $\pi_1,\ldots,\pi_m$ such that

$$\frac{\partial f}{\partial x_j} + \sum_{i=1}^{m} \pi_i \frac{\partial g_i}{\partial x_j} = 0 \qquad (j=1,\ldots,n)$$

$$g_i(\mathbf{x}) \leqslant b_i \qquad (i=1,\ldots,m)$$

$$\pi_i \geqslant 0 \qquad (i=1,\ldots,m)$$

$$\sum_{i=1}^{m} \pi_i(g_i(\mathbf{x}) - b_i) = 0$$

Actually this theorem requires a *constraint qualification* to exclude some pathological situations. Perhaps the easiest way to express this constraint qualification is to say that there must exist some point $\mathbf{x}$ where all the constraints are satisfied and $g_i(\mathbf{x})$ is strictly less than b_i for all i such that the function $g_i(\mathbf{x})$ is non-linear.

It is clear that the Kuhn–Tucker conditions are necessary for a local minimum even for problems lacking the convexity properties needed to ensure a global minimum.

10.3 SEPARABLE PROGRAMMING

Separable programming assumes that all the non-linear expressions in the problem can be separated out into sums and differences of non-linear functions of single arguments. This assumption may seem to severely restrict the usefulness of the method, but in fact a very large proportion of practical non-linear programming problems can conveniently be expressed in this form. We can handle problems of this type within a linear programming environment as follows.

Suppose that we have some variable z and that the function $f(z)$ occurs in either the objective function or a constraint. Suppose further that there are constants Z_0 and Z_I such that we know that

$$Z_0 \leqslant z \leqslant Z_I$$

Finally, suppose that Z_i, for $i = 0, 1, \ldots, I$, are an increasing sequence of numbers such that $f(z)$ is approximately linear between all the neighbouring points Z_i and Z_{i+1} (see Figure 10.3.1). Then, given such a *piecewise linear* approximation to $f(z)$, we can now introduce non-negative variables λ_i and the constraints

$$\sum_{i=0}^{J} \lambda_i = 1$$

$$\sum_{i=0}^{I} Z_i \lambda_i - z = 0$$

and we can approximate the function $f(z)$ by the linear function

$$\sum_{i=0}^{I} f(Z_i) \lambda_i$$

It is clear that the vertices of the piecewise linear approximation to $f(z)$ are obtained by putting all but one of the λ_i variables equal to zero. Any other point on the graph can be obtained by giving two neighbouring λ variables non-zero values and putting the others equal to zero.

Now if $f(z)$ is a convex function, any combination of non-neighbouring λ variables will overestimate it. Therefore there is no need to worry about invalid combinations occurring in the optimal solution to the problem if either $f(z)$ is

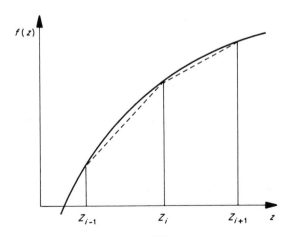

Figure 10.3.1

part of an objective function to be minimized or if $f(z)$ occurs on the left-hand side of a less-than-or-equal-to inequality.

When such convex problems occur we can approximate the function $f(z)$ by the piecewise linear function and then use the simplex method to solve the problem. In practice, we may have many different non-linear convex functions, each with different arguments, but this causes no difficulty.

If the above convexity conditions do not apply we can still insist that no invalid combinations of variables occur by using one of the following approaches:

(1) Separable programming, as defined by Miller (1963), finds a local optimum solution by restricting the choice of variable in the simplex method which is allowed to enter the basis. If one member of the set of variables denoted by λ_i is currently basic, then only its neighbours λ_{i-1} and λ_{i+1} are eligible to enter the basis. Also if two members of the set are currently basic then no other member may enter the basis until one of the existing members has left.

(2) An alternative approach finds a global optimum solution by a *branch and bound* technique. We first solve the problem as an ordinary linear programming problem, without imposing any of the restrictions about neighbouring λ variables. If the resulting solution satisfies the additional conditions on the λ variables, then all is well. Otherwise we choose some index, say k, and replace the original problem by two subproblems. In one subproblem we impose the additional requirement that

$$\lambda_0 = \lambda_1 = \cdots = \lambda_{k-1} = 0$$

and in the other subproblem we impose the additional requirement that

$$\lambda_{k+1} = \lambda_{k+2} = \cdots \lambda_I = 0$$

Clearly, any valid solution of the original problem must be a feasible solution of one or other of these subproblems. Therefore we now solve the subproblems, and if the optimum solution to either subproblem still fails to satisfy the conditions we can subdivide it further in the same way. Thus we build up a *branch and bound tree* which we have to search until we find the optimum solution to the problem. A fuller decription of these branch and bound techniques, and the different approaches which can be adopted to search the tree, is given in Chapter 11 on integer programming.

Therefore the final solution will not have more than two members of the set of λ variables with non-zero values, and if there are two such members they must be adjacent. Such a set, due to Beale and Tomlin (1970), is called a *special ordered set of type two*, or an *S2 set*.

A related situation arises if one has a set of variables of which, at most, one may be non-zero. Beale and Tomlin call such a set a *special ordered set of type one*, or *S1 set*. In this context they would be appropriate if z was required to be equal to Z_i for some i. For example, if z represents the diameter of a pipe, then we might reasonably require it to be equal to a value for which pipe is manufactured. If a trial solution includes two or more members of a set we can then subdivide the problem by noting that

either
$$\lambda_0 = \lambda_1 \cdots = \lambda_k = 0$$

or
$$\lambda_{k+1} = \lambda_{k+2} = \cdots = \lambda_I = 0$$

An alternative form of separable programming provided in some mathematical programming systems, generally called δ-*separable*, replaces the variables λ_i by variables δ_i that are restricted to be between 0 and 1 such that $\delta_i = \lambda_i - \lambda_{i-1}$ for $i = 1, \ldots, I$. We then have

$$\sum_{i=1}^{I} (Z_i - Z_{i-1})\delta_i - z = Z_0$$

and $f(z)$ is approximated by the linear function

$$f(Z_0) + \sum_{i=1}^{I} (f(Z_i) - f(Z_{i-1}))\delta_i$$

The simplex method basis entry rules for δ-separable on non-convex problems are that δ_i can only enter the basis from its lower bound if δ_{i-1} is out of the basis at its upper one, while δ_i can only enter the basis from its upper bound if δ_{i+1} is out of the basis at its lower bound.

10.4   EXAMPLE OF SEPARABLE PROGRAMMING

These ideas can be illustrated in an admittedly somewhat artificial context by considering Hartman's family of test problems for global optimization. Suppose that we have to devise a procedure which can be used to find the global maximum of the function

$$\sum_{i=1}^{m} C_i \exp\left(- \sum_{j=1}^{n} A_{ij}(x_j - P_{ij})^2 \right)$$

with respect to the decision variables $x_1, \ldots, x_n$, where each variable is required to lie between 0 and 1 and where m, n, C_i, A_{ij} and P_{ij} are all given data. We assume that all C_i and A_{ij} are non-negative. This is one of the test functions considered by Dixon and Szegö (1978).

A formulation using S2 sets can be developed by writing

$$f(\theta) = \theta^2, \quad \text{and} \quad g(\theta) = \exp(-\theta)$$

Then the problem is to choose **x**, **y** and **z** to maximize

$$\sum_{i=1}^{m} C_i g(z_i)$$

subject to the constraints

$$z_i - \sum_{j=1}^{n} A_{ij}(y_j - 2P_{ij}x_j + P_{ij}^2) = 0 \qquad (i = 1, \ldots, m)$$

$$y_j - f(x_j) = 0 \qquad (j = 1, \ldots, n)$$

Since $0 \leqslant x_j \leqslant 1$ we note that

$$Z_{\text{MIN}i} \leqslant z_i \leqslant Z_{\text{MAX}i} \qquad (i = 1, \ldots, m)$$

where

$$Z_{\text{MIN}i} = \sum_{j=1}^{n} A_{ij}(\max(0, -P_{ij}, P_{ij} - 1))^2$$

and

$$Z_{\text{MAX}i} = \sum_{j=1}^{n} A_{ij} \max(P_{ij}^2, (1 - P_{ij})^2)$$

The complete formulation is then to choose non-negative $\mathbf{x}, \mathbf{y}, \lambda$ and $\boldsymbol{\mu}$ to maximize x_0 subject to

$$x_0 - \sum_{i=1}^{m} \sum_{k} C_i \exp(-Z_{ik})\mu_{ik} \qquad\qquad = 0$$

$$\sum_{k} \mu_{ik} \qquad\qquad\qquad = 1 \qquad\qquad (i = 1, \ldots, m)$$

$$\sum_{k} Z_{ik}\mu_{ik} - \sum_{j} A_{ij}y_j + 2\sum_{j} A_{ij}P_{ij}x_j = \sum_{j} A_{ij}P_{ij}^2 \qquad (i = 1, \ldots, m)$$

$$y_j - \sum_{k} X_{jk}^2 \lambda_{jk} \quad = 0 \qquad\qquad (j = 1, \ldots, n)$$

$$\sum_{k} \lambda_{jk} \quad = 1 \qquad\qquad (j = 1, \ldots, n)$$

$$-x_j + \sum_{k} X_{jk}\lambda_{jk} \quad = 0 \qquad\qquad (j = 1, \ldots, n)$$

This problem has $m + n$ S2 sets and $2m + 3n$ constraints. There is no temptation to overestimate y_j, so we never need to branch on any of the λ-sets, but we will have to branch on the m μ-sets.

10.5   REDUCED GRADIENT METHODS

The other general approach that has been used successfully on large-scale non-linear programming can be traced back to ideas on a reduced gradient method suggested by Wolfe (1963). This approach was generalized to non-linear constraints by Abadie and Carpentier (1969) and another implementation is described by Murtagh and Saunders (1978).

This approach finds a possibly local optimum solution to a problem containing arbitrary differentiable objective and constraint functions. The disadvantage compared with methods based on separable programming is that the optimization algorithm is interrupted at each iteration to collect new data on the function values and first derivatives of the objective and constraint functions at the current trial solution. The approach cannot be extended to provide a guaranteed optimum to a non-convex problem. However, it does not need a prior analysis of the mathematical structure of the non-linear functions, and is more convenient when any of these functions cannot easily be expressed in separable form. The approach is a generalization of the simplex method for linear programming on the following lines.

First write the problem as one of minimizing a function $f(\mathbf{x})$ subject to equality constraints

$$g_i(\mathbf{x}) = b_i \qquad (i = 1, \ldots, m)$$

and simple lower and upper bounds on the individual variables

$$L_j \leqslant x_j \leqslant U_j \qquad (j = 1, \ldots, n)$$

This can easily be done by introducing slack variables. In practice, many $L_j = 0$ and many $U_j = \infty$.

Then, given a feasible trial solution, make local linear approximations to all the constraints and solve these equations for some variables, the basic ones, in terms of the others. As far as possible, choose the basic variables from among those that lie strictly between their lower and upper bounds. We assume in this outline explanation that this can always be done. We may write the solved form of these equations in the conventional tableau notation

$$X_i = \bar{a}_{i0} - \sum_j \bar{a}_{ij} x_j \quad (i = 1, \ldots, m) \tag{10.5.1}$$

where summation extends over all non-basic variables.

We now need to know whether the current trial solution is optimum, and if not, how to improve it. We assume that we can calculate the first derivatives $\partial f / \partial x_j$ of the objective function with respect to all the variables. If we now consider moving a small distance $\Delta \mathbf{x}$ from the current trial solution, $\mathbf{a}$ say, a first-order Taylor expansion gives us

$$f(\mathbf{a} + \Delta \mathbf{x}) \simeq f(\mathbf{a}) + \mathbf{c}^T \Delta \mathbf{x}$$

where $\mathbf{c}$ is the vector of first partial derivatives evaluated at the trial solution $\mathbf{a}$. However, these partial derivatives are not immediately relevant, since what we really need to know are the partial derivatives of the objective function with respect to the non-basic variables when the basic variables are adjusted so that the constraints remain satisfied. Rewriting $\mathbf{c}^T \Delta \mathbf{x}$ in terms of the basic and non-basic variables we obtain

$$\sum_{i \, \text{basic}} \frac{\partial f}{\partial X_i} \Delta X_i + \sum_{j \, \text{non-basic}} \frac{\partial f}{\partial x_j} \Delta x_j \tag{10.5.2}$$

where the partial derivatives are all evaluated at the trial solution $\mathbf{a}$. However, from equation (10.5.1) we see that

$$\Delta X_i = - \sum_{j \, \text{non-basic}} \bar{a}_{ij} \Delta x_j$$

so that equation (10.5.2) reduces to

$$\sum_{j \, \text{non-basic}} \left[\frac{\partial f}{\partial x_j} - \sum_{i \, \text{basic}} \bar{a}_{ij} \frac{\partial f}{\partial X_i} \right] \Delta x_j$$

so we have, to first order,

$$f(\mathbf{a} + \Delta\mathbf{x}) \simeq f(\mathbf{a}) + \sum_{j\,\text{non-basic}} d_j \Delta x_j$$

where

$$d_j = \frac{\partial f}{\partial x_j} - \sum_{i\,\text{basic}} \bar{a}_{ij} \frac{\partial f}{\partial X_i}$$

evaluated at the current trial solution. These required partial derivatives d_j are known as the *reduced costs*, or alternatively as the components of the reduced gradient.

The first-order optimality conditions are then satisfied at the current trial solution if, for all j, either

$$x_j = L_j \qquad \text{and} \quad d_j \geqslant 0$$

or

$$L_j < x_j < U_j \quad \text{and} \quad d_j = 0$$

or

$$x_j = U_j \qquad \text{and} \quad d_j \leqslant 0$$

If these conditions are not all satisfied, then we must start an unconstrained optimization routine in the space of those non-basic variables that lie between their bounds and perhaps one or more of the variables that can profitably leave either their lower or upper bounds. The variables over which this unconstrained optimization takes place may be called the *independent* variables. (Abadie and Carpentier, 1969, use the symbol I to denote the set of these variables. The term *superbasic* variables is used by Murtagh and Saunders, 1978, and this is logical to the extent that the division of variables between the basic and independent variables is often arbitrary. However, I do not like this terminology since the independent variables are logically non-basic.)

This process continues until one of the independent variables hits a bound, in which case it will be held at this bound and further optimization continues without it, or until a basic variable hits a bound, in which case it is made non-basic in place of one of the independent variables.

To illustrate the method, consider choosing non-negative p_1, p_2 and p_3 to minimize

$$f(p_1, p_2, p_3) = 1 + p_2 - 5p_1 p_2 - 2p_1 p_3 - p_2 p_3$$

subject to

$$p_1 + p_2 + p_3 = 1$$

starting at the point $(0, 0, 1)$. The first partial derivatives are

$$\frac{\partial f}{\partial p_1} = -5p_2 - 2p_3$$

$$\frac{\partial f}{\partial p_2} = 1 - 5p_1 - p_3$$

$$\frac{\partial f}{\partial p_3} = -2p_1 - p_2$$

We begin by making p_3 a dependent variable, i.e.

$$p_3 = 1 - p_1 - p_2$$

so

$$d_1 = \frac{\partial f}{\partial p_1} - \frac{\partial f}{\partial p_3} \qquad d_2 = \frac{\partial f}{\partial p_2} - \frac{\partial f}{\partial p_3}$$

Therefore the reduced costs d_1 and d_2 evaluated at the point $(0, 0, 1)$ are -2 and 0, respectively. We can therefore profitably increase p_1.

We begin by making a unit increase to assess the rate of change of the gradient vector. Calculating the first partial derivatives and the reduced costs at the feasible point $(1, 0, 0)$ we obtain

$$\frac{\partial f}{\partial p_1} = 0, \quad \frac{\partial f}{\partial p_2} = -4, \quad \frac{\partial f}{\partial p_3} = -2, \quad d_1 = 2, \quad d_2 = -2$$

To make $d_1 = 0$ we must reduce the step length to $\frac{1}{2}$, so we move to the feasible point $(\frac{1}{2}, 0, \frac{1}{2})$, where $d_1 = 0$ and $d_2 = -1$.

The independent variable p_1 is now between its bounds with a zero reduced cost. However, d_2 is now negative, so we increase p_2 keeping p_1 non-negative. Our embedded unconstrained optimization is in two dimensions. It is desirable to move in a direction in which d_1 remains equal to zero. This means moving in a direction *conjugate* to the previous search direction (see Section 4.3 for a description of conjugate directions). The gradient difference vector is

$$\begin{pmatrix} 2 \\ -2 \end{pmatrix} - \begin{pmatrix} -2 \\ 0 \end{pmatrix} = \begin{pmatrix} 4 \\ -2 \end{pmatrix}$$

To produce a search direction ξ_2 conjugate to $\binom{1}{0}$ we write $\xi_2 = \binom{0}{1} + b\binom{1}{0}$ where

$$b = -\frac{\begin{pmatrix} 0 \\ 1 \end{pmatrix}^T \begin{pmatrix} 4 \\ -2 \end{pmatrix}}{\begin{pmatrix} 1 \\ 0 \end{pmatrix}^T \begin{pmatrix} 4 \\ -2 \end{pmatrix}} = \frac{1}{2}$$

Therefore the new search direction ξ_2 is $\begin{pmatrix} \frac{1}{2} \\ 1 \end{pmatrix}$. We make a unit increase in this direction to assess the rate of change of the gradient vector. So we move to the point $(1, 1, -1)$, where

$$\frac{\partial f}{\partial p_1} = -3, \quad \frac{\partial f}{\partial p_2} = -3, \quad \frac{\partial f}{\partial p_3} = -3, \quad d_1 = 0, \quad d_2 = 0$$

However, this point is infeasible, and we can only move a distance of $\frac{1}{3}$, to $(\frac{2}{3}, \frac{1}{3}, 0)$. The basic variable p_3 has hit its lower bound of zero, so we make it independent. Let us write

$$p_2 = 1 - p_1 - p_3$$

Then the reduced costs d_1 and d_3 are given by

$$d_1 = \frac{\partial f}{\partial p_1} - \frac{\partial f}{\partial p_2}, \quad d_3 = \frac{\partial f}{\partial p_3} - \frac{\partial f}{\partial p_2}$$

and at the point $(\frac{2}{3}, \frac{1}{3}, 0)$ they both have the value $\frac{2}{3}$. However, the solution is still not optimal, because p_1 is between its bounds with a non-zero reduced cost, so we decrease p_1. Making a unit decrease to assess the rate of change of the gradient vector, we find that at the point $(-\frac{1}{3}, \frac{4}{3}, 0)$

$$\frac{\partial f}{\partial p_1} = \frac{-20}{3}, \quad \frac{\partial f}{\partial p_2} = \frac{8}{3}, \quad \frac{\partial f}{\partial p_3} = \frac{-2}{3}, \quad d_1 = \frac{-28}{3}, \quad d_3 = \frac{-10}{3}$$

To make $d_1 = 0$ we reduce the step length to $\frac{1}{15}$ and move to the point $(\frac{3}{5}, \frac{2}{5}, 0)$, where $d_1 = 0$ and $d_3 = \frac{2}{5}$. This is the optimum solution.

11

Integer Programming

11.1 INTRODUCTION

Integer programming, sometimes called integer linear programming (ILP) or mixed integer programming (MIP), refers to linear programming with the additional restriction that some or all of the variables are required to take integer values. In practice, these variables are often *zero–one* variables—variables which can only take the value 1 or 0—indicating whether or not some activity should take place. However, the solution methods are equally applicable to integer variables that can take any non-negative integer values.

Integer programming is now an established tool in practical operational research, being applied to many problems involving set-up costs, fixed costs and other economies of scale. Until a few years ago these applications were largely confined to medium- and long-term capital investment projects where computing costs were of secondary importance. The technique is now both cheap and reliable enough to be used for various weekly or even daily production scheduling tasks where a purely linear programming model is unrealistic. On the other hand, the technique must be handled carefully, since some integer programming formulations can consume a large amount of computer time without producing any useful results. Therefore, given a new type of problem, with more than, say, 50 integer variables, one cannot be confident that integer programming will work until it has been tried on a realistic set of test data. If it does not work, then an alternative mathematical formulation of the same physical problem may work better. Two principles can guide us in finding a suitable alternative formulation. One is that we should try to reduce the discrepancy between the true (integer) solution to the problem and the continuous solution, i.e. the solution to the relaxed problem without the requirement that certain variables must be integer. The other principle is that an effective integer programming model should usually have a single clear purpose. If this is to choose which *combination* of n major investments to select (see Section 1.2) then the consequences of selecting any individual investment should be represented simply, and one should generally avoid other integer variables representing alternative minor choices consequential on each major choice. This may require a detailed analysis of each investment before formulating the integer programming model. However, one can generally afford to do n such detailed analyses if this simplifies the task of picking the best of the 2^n possible combinations of investments.

It is worth spending a few minutes on the general problem of how to find the best combination of n choices, since the straightforward approach of evaluating them all and picking the best is a classical example of an algorithm in which the computing task grows exponentially with the problem size.

If $n = 10$, $2^n \simeq 10^3$, and it may be fairly easy to assess each possibility and pick out the best. If $n = 20$, $2^n \simeq 10^6$, and the task has become more formidable. If $n = 30$, $2^n \simeq 10^9$. Straight enumeration is now almost certainly impossible. Note, in particular, that the 30-variable problem is a thousand times more difficult than the 20-variable one. Note also that some clever scheme that saved only 99% of the work of complete enumeration would be of only limited value.

The problem of finding the best *permutation* of n objects is even worse. The number of permutations is $n!$. This is about 3.6×10^6 if $n = 10$, which may be near the limit of feasibility for complete enumeration. However, the work goes up by more than a factor of 1000 if n increases from 10 to 13.

The potential difficulties of integer programming are therefore formidable. On the other hand, the scope of integer programming is much wider than may be immediately apparent. In particular, Markowitz and Manne (1957) showed that integer programming could be applied to the task of finding global optimum solutions to non-convex problems on continuous variables. Extensions of this idea were covered in Chapter 10 on non-linear programming.

11.2 BRANCH AND BOUND METHODS

From the point of view of numerical analysis, the most important thing about integer programming is that the conventional strategy of 'hill-climbing' for numerical optimization will not work. This strategy is to take a trial solution, see if any small change improves it and if so, modify the trial solution in the indicated direction. If not, we have achieved a local optimum solution and can stop. The simplex method for linear programming is a hill-climbing method, and we know from mathematical theory that a local optimum is then also a global optimum. However, integer programming problems typically have many local optima that are much worse than the global optimum. So what are we to do? Most methods for solving integer programming problems are based on the following general strategy.

We wish to maximize $f(\mathbf{x})$ subject to the constraint that $\mathbf{x}$ lies in some region R. Suppose that we can find a point $\mathbf{x}^1$ that maximizes $f(\mathbf{x})$ subject to the constraint that $\mathbf{x}$ lies in the region R_1, where R_1 also includes the original region R. This is much easier than the original problem if $f(\mathbf{x})$ is concave and the region R_1 is convex, but R is not convex—since we can then use a hill-climbing method to find a local optimum which will necessarily be a global optimum. The problem is not much more difficult if R_1 consists of the union of a moderate number of convex sets, since we simply have to find the optimum in each set and take the largest.

If the point $\mathbf{x}^1$ lies in the region R we have solved the original problem.

Otherwise we can modify R_1 so that it excludes the point $\mathbf{x}^1$, but no other point in the region R. We can now solve the modified problem. More generally, if we know a point $\mathbf{x}^0$ which lies in the region R, known as the *incumbent solution*, then we need only insist that the modified R_1 does not exclude any points in R for which $f(\mathbf{x}) > f(\mathbf{x}^0)$. Whenever we find a point in R with $f(\mathbf{x}) > f(\mathbf{x}^0)$ this becomes the new incumbent solution. This process is repeated until we have a solution to the modified problem that is in R (a new incumbent solution), or until R_1 is empty because there are no more points in R with $f(\mathbf{x}) > f(\mathbf{x}^0)$.

The original region R_1 is usually taken as the set of points satisfying all the equality and inequality constraints, but omitting the requirement that the variables must take integer values. Therefore for a normal integer programming problem the task of minimizing $f(\mathbf{x})$ subject to the constraint that $\mathbf{x}$ lies in the region R_1 is just a linear programming problem, which we can solve using the simplex method. The modifications to R_1 then usually consist of adding linear inequality constraints.

If a single inequality is added it is known as a *cutting plane*. This approach was pioneered by Gomory (1958), and many alternative types of cutting plane have been developed. However, the apparently more pedestrian approach of adding two alternative linear inequalities has proved more reliable. This was called *branch and bound* by Little *et al.* (1963). This means that the original linear programming problem is replaced by two linear programming subproblems, and at subsequent steps of the algorithm one subproblem is replaced by two new ones unless its optimum solution is either in R or has a value worse than $f(\mathbf{x}^0)$.

In practice the inequalities added are usually more restrictive upper or lower

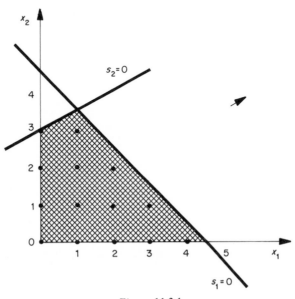

Figure 11.2.1

bounds on the integer variables. For example, let us consider the simple problem illustrated in Figure 11.2.1:

maximize $\qquad\qquad\qquad 5x_1 + 2x_2$

subject to

$$x_1 + x_2 \leqslant 4.5$$
$$-x_1 + 2x_2 \leqslant 6.0$$
$$x_1, x_2 \geqslant 0, \text{ integer}$$

Solving the linear problem (in other words, neglecting the fact that x_1 and x_2 have to take integer values) we have the tableau:

$$x_0 = 5x_1 + 2x_2$$
$$s_1 = 4.5 - x_1 - x_2$$
$$s_2 = 6.0 + x_1 - 2x_2$$

Making x_1 basic and s_1 non-basic:

$$x_0 = 22.5 - 5s_1 - 3x_2$$
$$x_1 = 4.5 - s_1 - x_2$$
$$s_2 = 10.5 - s_1 - 3x_2$$

so the optimum solution to the relaxed problem is at $(4.5, 0)$ with the objective value of 22.5.

However, x_1 is not an integer value, so we *branch* on the variable x_1 and create two subproblems, one with the additional constraint that $x_1 \leqslant 4$ and the other with the additional constraint that $x_1 \geqslant 5$. Clearly, the second subproblem is infeasible, so we solve the first subproblem, treating the upper bound on x_1 as an explicit constraint for simplicity:

$$x_0 = 5x_1 + 2x_2$$
$$s_1 = 4.5 - x_1 - x_2$$
$$s_2 = 6.0 + x_1 - 2x_2$$
$$s_3 = 4.0 - x_1$$

Making x_1 basic and s_3 non-basic we obtain

$$x_0 = 20.0 - 5s_3 + 2x_2$$
$$s_1 = 0.5 + s_3 - x_2$$
$$s_2 = 10.0 - s_3 - 2x_2$$
$$x_1 = 4.0 - s_3$$

If we now make x_2 basic and s_1 non-basic the tableau becomes

$$x_0 = 21.0 - 3s_3 - 2s_1$$
$$x_2 = 0.5 + s_3 - s_1$$
$$s_2 = 9.0 - 3s_3 + 2s_1$$
$$x_1 = 4.0 - s_3$$

so the optimum solution to this subproblem is at the point $(4, 0.5)$ with the objective value of 21.0.

Note that the objective value has decreased compared with the solution of the relaxed problem. This is to be expected, because any solution to the subproblem is also a solution to the parent problem, so the objective value cannot increase above the optimum of the parent problem. However, in the optimum solution to the previous subproblem x_2 is not an integer value. Therefore we branch on the variable x_2 and create two further subproblems, one with the additional constraint that $x_2 \leqslant 0$ and the other with the additional constraint that $x_2 \geqslant 1$.

In the first of these subproblems x_2 is obviously fixed at the value zero so, by inspection, we can easily see that the optimum solution to this subproblem will be at the point $(4, 0)$ with the value 20.0. This solution clearly satisfies all our integer constraints and is a solution to our original problem. However, we cannot be certain that it is the *global* solution until we have solved the second subproblem and found all the possible integer solutions. Thus solving the second subproblem, treating the lower bound on x_2 as an explicit constraint, we obtain the tableau:

$$x_0 = \quad\quad 5x_1 + 2x_2$$
$$s_1 = 4.5 - \quad x_1 - \quad x_2$$
$$s_2 = 6.0 + \quad x_1 - 2x_2$$
$$s_3 = 4.0 - \quad x_1$$
$$A_1 = 1.0 \quad\quad\quad - \quad x_2 + s_4$$

Making x_2 basic and A_1 non-basic:

$$x_0 = 2.0 + 5x_1 - 2A_1 + 2s_4$$
$$s_1 = 3.5 - \quad x_1 + \quad A_1 - \quad s_4$$
$$s_2 = 4.0 + \quad x_1 + 2A_1 - 2s_4$$
$$s_3 = 4.0 - \quad x_1$$
$$x_2 = 1.0 \quad\quad\quad - \quad A_1 + \quad s_4$$

Making x_1 basic and s_1 non-basic:

$$x_0 = 19.5 - 5s_1 + 3A_1 - 3s_4$$
$$x_1 = \quad 3.5 - \quad s_1 + \quad A_1 - \quad s_4$$
$$s_2 = \quad 7.5 - \quad s_1 + 3A_1 - 3s_4$$
$$s_3 = \quad 0.5 + \quad s_1 - \quad A_1 + \quad s_4$$
$$x_2 = \quad 1.0 \quad\quad\quad - \quad A_1 + \quad s_4$$

which is optimum with the value 19.5. The value of x_1 in the solution to this subproblem is a non-integer value so we may think that we need to branch further on the variable x_1. However, this is not necessary, because the objective value we just obtained is less than that of the integer solution with the value of 20.0. Therefore because the objective values can only worsen as we impose more constraints we know that the integer solution $(4, 0)$ with the value of 20.0 must be the global solution of the original problem.

In the previous example we treated the imposed bounds as explicit constraints so that the size of the linear programming subproblems grew as we branched. In practice, however, upper and lower bounds are treated implicitly (see Section 6.3) so that the size of the subproblems remains constant. Furthermore, we solved each subproblem by beginning from an *all-slack* basis, but the only differences between the subproblems are in the lower and upper bounds on the integer variables. Therefore in practice we use the information provided by the optimal basis of the parent problem to begin iterating, since this is likely to be a better trial solution than starting from an all-slack basis again.

For each subproblem we store:

(1) The lower and upper bound for each integer variable;
(2) The optimum basis for the linear programming subproblem from which the new subproblem was generated;
(3) A guaranteed upper bound U on the *value* of the subproblem, where the value is defined as the maximum value of $f(\mathbf{x})$ for any $\mathbf{x}$ in the region R that also satisfies the additional constraints of the subproblem;
(4) Some estimate E of this value.

This procedure is often represented as a *tree search*. Each node of the tree represents a subproblem; the root represents the original linear programming

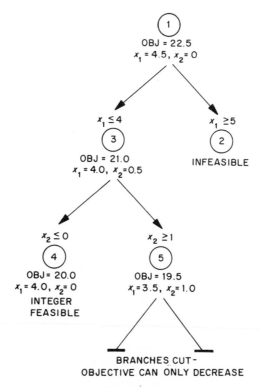

Figure 11.2.2

problem; and when any node or subproblem is explored, branches are drawn from this node to the nodes representing any subproblems generated from it. The list of alternative subproblems to be solved is therefore called the *list of unexplored nodes*. Figure 11.2.2 illustrates the branch and bound tree for the earlier example, and this tree-structure representation clarifies the logic of the method. Fortunately it does not need to be stored explicitly in the computer.

This general description leaves five important questions unresolved:

(1) Which subproblem should be explored next?
(2) How should it be solved?
(3) On which variable should we branch, assuming that the subproblems are generated by tightening the bounds on a single variable?
(4) How should U, the guaranteed upper bound on the value of the subproblem, be computed? (A simple solution is to use the value of the relaxed problem, but it may be possible to refine this.)
(5) How should E, the estimate of the value of the subproblem, be computed?

There are various approaches which can be used in answer to these problems, but we will only concern ourselves with the first question. There is a theoretical agrument for always exploring the subproblems with the highest value of U, because this would minimize the number of subproblems created before completing the search if the branching procedure from this subproblem were unaffected by the value $f(\mathbf{x}^0)$ of the incumbent solution. However, this strategy makes no attempt to find any good integer solutions early in the search, and is subject to five objections:

(1) The branching procedure can sometimes be improved if we have a good value of the incumbent solution $f(\mathbf{x}^0)$.
(2) A good value of $f(\mathbf{x}^0)$ reduces the work on some subproblems, since they can be abandoned before they are fully solved if it can be shown that the linear programming optimum is no better than $f(\mathbf{x}^0)$.
(3) We can often arrange to solve either, or even both, of the subproblems that have just been generated without re-inverting to the current basis.
(4) The list of unexplored nodes may grow very long and create computer storage problems.
(5) No useful results may be obtained if the search has to be ended before a guaranteed optimum has been found.

An alternative strategy was advocated by Little *et al.* (1963) and was used in early work on large-scale integer programming (see Beale, 1968). This is *always branch right* or, equivalently, *last in first out*. It minimizes the number of subproblems in the list at any time, and usually finds a good solution early, which can be used to improve the rest of the search. It also allows the subproblems in the list to be stored in a sequential store such as magnetic tape. However, it has been found inadequate on larger and more difficult problems, and has been discarded.

There is not much wrong with 'last in first out' as long as we are branching on one of the subproblems just created, and this is particularly true if the more promising of these subproblems is always chosen. However, when neither subproblem leads to any new subproblem, 'last in first out' involves bracktracking through the more recently created subproblems, and this is often unrewarding.

Practical integer programming codes now contain many options, and it is usual to adopt a compromise between 'last in first out' and exploring the subproblems with the highest value of U. A more detailed discussion of this and the other questions is contained in Beale (1977).

11.3 FORMULATION OF INTEGER PROGRAMMING PROBLEMS

The formulation of integer programming problems is a somewhat delicate art. As noted earlier, success with large-scale integer programming models usually depends on having a formulation in which the solution of the relaxed problem gives good guidance about the form of the optimum integer solution.

The basic way to represent a yes-or-no decision in a mathematical model is to use a *zero–one* or *binary* variable. The formulation is then given by

$$x - M\delta \leqslant 0$$

where δ is a zero–one variable and x is a non-negative variable that is not allowed to be non-zero unless $\delta = 1$, with a natural bound of M. If there is a fixed cost F that is incurred whenever x is non-zero, we can then associate this cost with the zero–one δ variable. This is a very powerful device that can be extended in various ways, but it is important to choose values of M which are as small as justifiably possible to speed up the branch and bound search procedure.

An example of a formulation which uses a zero–one variable is the simple plant-location problem. A company needs to know the best places to locate its industrial plants so that it can supply all its customers at the cheapest cost. This cost includes both the transportation costs and the fixed costs associated with each plant. The natural formulation is

Subscripts

c	Customer area
l	Possible location

Constants

C_{Dcl}	Total cost of delivering to customer area c from location l (£/tonne)
C_{Fl}	Fixed cost of having a plant in location l (thousands of £)
R_c	Total requirements in customer area c (thousands of tonnes)

Variables

δ_l Zero–one variable, defining whether or not plant should be built at location l

x_{cl} Fraction of demand in customer area c to be supplied from location l

The problem is to minimize the sum of the fixed costs of the plants plus the sum of the costs of meeting the demands in all customer areas:

minimize

$$\sum_l C_{Fl}\delta_l + \sum_c \sum_l C_{Dcl}R_c x_{cl}$$

subject to the constraint that the demand in each customer area must be met:

$$\sum_l x_{cl} = 1 \qquad \text{(for all } c\text{)}$$

where
$$x_{cl} = 0 \qquad \text{if } \delta_l \leqslant 0$$

The last constraint can be written most compactly in the form

$$\sum_c x_{cl} - M_l \delta_l \leqslant 0 \qquad \text{(for all } l\text{)}$$

where M_l denotes the maximum number of customer areas that location l can possibly supply. This might be C, the total number of customer areas. Although this formulation is very compact because it has the least number of constraints, branch and bound is very inefficient at solving the problem because the integer solution and the continuous or relaxed solution are very different (the continuous solution will allow a plant to be fractionally built at different sites). Therefore, in spite of the great increase in the number of constraints, it is better to replace these constraints by

$$x_{cl} - \delta_l \leqslant 0 \qquad \text{(for all } c, l\text{)}$$

This is typical of integer programming problems—the time spent in branch and bound is often dramatically reduced by adding more constraints to the problem, providing these constraints bring the continuous solution closer to the integer solution.

12

Dynamic Programming

12.1 INTRODUCTION

Dynamic programming is an alternative to integer programming for some problems in combinatorial optimization. Its scope is much more restricted, but when it is applicable it is sometimes much more efficient than integer programming, and the reason for this is illuminating.

Most applications of dynamic programming to deterministic problems can be expressed as problems of finding the *shortest route* through a network of directed arcs, whose lengths may be either positive, zero or even negative since the lengths typically represent costs. Now if we are finding the shortest route from A to B we may have to consider going through X or Y or Z. However, if we go through X then the best route from A to X will not depend on how we choose to go from X to B. Dynamic programming exploits this important fact, but current branch and bound methods for integer programming do not. This may not be too serious if we are prepared to accept a good solution that is not a guaranteed optimum, but it can add enormously to the task of completing the integer programming tree search.

12.2 SHORTEST-ROUTE PROBLEMS

The art of dynamic programming is essentially one of formulating a problem as a shortest-route problem. Before illustrating this it is appropriate to discuss the solution of such problems. There are a number of variants of the solution algorithm, and the best one to use depends on the nature of the data, but the point is that the problem is in principle an easy one.

The network is defined as a set of directed arcs. Each has a start node N_i, an end node N_j and a *generalized length* or *cost* a_{ij}, which may be positive, or zero or even negative. We are interested in finding the shortest route, and its total generalized length, from some origin node N_0 to some finishing node N_F. Figure 12.2.1 is an example of such a network. The basic algorithm is as follows.

Let D_j denote the minimum generalized length of any route considered so far from the origin to node N_j, and let P_j denote the *predecessor* to node j on this route, i.e. P_j is the last node visited before node N_j. Initially, set $D_0 = 0$ and

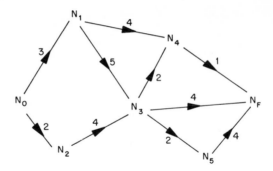

Figure 12.2.1

$D_j = \infty$ for $j \neq 0$. Then consider each arc in turn, and if $D_j > D_i + a_{ij}$, replace D_j by $D_i + a_{ij}$ and set $P_j = N_i$.

Cycle through all the arcs in this way until a *complete cycle* has been made without changing any D_j. The final value of D_F is then the required minimum generalized length, and the corresponding route can be traced back from the finishing node. The last node before N_F is P_F, the last node before this is the predecessor of P_F and so on, until we find a node whose predecessor is N_0.

For example, in the network illustrated in Figure 12.2.1, the values of D_j and P_j are

$$D_1 = 3, \quad P_1 = N_0$$
$$D_2 = 2, \quad P_2 = N_0$$
$$D_3 = 6, \quad P_3 = N_2$$
$$D_4 = 7, \quad P_4 = N_1$$
$$D_5 = 8, \quad P_5 = N_3$$
$$D_F = 8, \quad P_F = N_4$$

so that the shortest route through the network is of length 8 and goes from N_0 to N_1 to N_4 to N_F.

This basic algorithm may be slow, and it may even cycle endlessly: it will do so if and only if the network contains any cycles (or loops) of negative total length. Fortunately, in most dynamic programming applications we do not have to consider the possibility of such cycles. Indeed in most applications the nodes are grouped into *stages* such that each arc leads from a node in one stage to a node in the next one. Therefore the network looks somewhat like the one in Figure 12.2.2.

If we now give the nodes two subscripts, say i and t, where the second subscript represents the stage, and let a_{ijt} denote the length of the arc from node N_{it-1} to node N_{jt}, then the algorithm can be rewritten as

$$D_{jt} = \min_i (a_{ijt} + D_{it-1})$$

and P_{jt} is the value of i corresponding to the node N_{it-1} for which this minimum is attained.

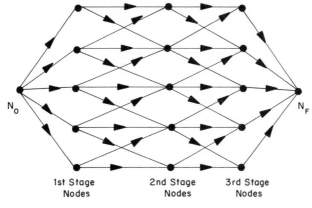

1st Stage   2nd Stage   3rd Stage
Nodes   Nodes   Nodes

Figure 12.2.2

Note that as far as the second stage this is just a systematic way of examining all possibilities and choosing the best, but at the third stage there is a genuine saving in work. For example, labelling the nodes at stage t as A_t, B_t, C_t, D_t and E_t, if the path

$N_0 A_1 B_2 A_3 N_F$

is shorter than

$N_0 B_1 B_2 A_3 N_F$

then we do not need to consider the path

$N_0 B_1 B_2 B_3 N_F$

because

$N_0 A_1 B_2 B_3 N_F$

must be shorter, and the above algorithm exploits this fact.

An equivalent algorithm can be used to compute the shortest distance from each node to the finishing node. If V_{it} denotes this distance, then the algorithm is

$$V_{it} = \min_j (a_{ijt+1} + V_{jt+1})$$

We then need to store S_{it}, the successor node to N_{it}, if we pass through this node. This *backwards recursion* formula is essentially a statement of Bellman's (1957) *principle of optimality*.

In practice, it is usually slightly better to use the original forwards recursion formula on deterministic problems, because there are usually fewer nodes that can be reached in the first few stages from the original than there are nodes in the final stages from which one can reach the finishing node. However, the backwards recursion formula can also be applied to some *stochastic* problems. If we are at node N_{it} and take a decision j, then in the deterministic problem we reach node N_{jt+1} with probability 1, having incurred a cost a_{ijt+1}. However, if instead we have a probability p_{ijkt} of reaching node N_{kt+1}, then we can apply the formula

$$V_{it} = \min_j \left(a_{ijt+1} + \sum_k p_{ijkt} V_{kt+1} \right)$$

where V_{it} denotes the expected total cost of progressing from node N_{it} to the finishing node. This stochastic problem looks very general, but note that the probability of moving from node N_{it} to node N_{kt+1} is assumed to depend only on the current decision j and not on how we reached the node N_{it}.

An apparently less general formulation of stochastic dynamic programming that is more useful in practice is as follows. The problem is still concerned with finding a route through a network from an origin node to a finishing node. Each node is *either* a choice node, where one must choose which outgoing arc to take, *or* a chance node, where the choice is made at random but with known probabilities. The task is then to devise a strategy for the choice at each choice node so as to minimize the mean (or expected) value of the length of the path to the finishing node. In most applications the arcs starting at choice nodes all end at chance nodes, and the arcs starting at chance nodes all end at choice nodes (or the finishing node). If at some stage the nodes require both a choice and a random effect, the model can be revised so that this stage is divided into two—a stage containing only choice nodes followed by one containing only chance nodes. This reduces the computing effort if the chance nodes can be reached from more than one choice node. The solution is found by backward recursion from the finishing node.

12.3   AN EXAMPLE PROBLEM

As we mentioned earlier, the art of dynamic programming is not finding the route through the network but formulating the problem as a shortest-route one in the first place. Typical problems which can be solved using dynamic programming techniques require plans to be made for a sequence of time periods. The following is a simple example.

We have to plan the development of a new industrial plant to manufacture and sell a product. We know forecasts D_t (for $t = 1, \ldots, T$) of the demand (in tonnes/year) for this product for each of the next T years and we also know the profit P (revenue less variable operating costs) achieved by selling the product (in £/tonne). Suppose further that the total capital plus fixed operating costs (until the end of the time horizon) incurred in year t when the capacity of the plant is increased by x tonnes is some known function $C_t(x)$. Then the problem can be formulated as a non-convex, non-linear programming one as follows.

Variables

x_t increase in capacity in year t (tonnes/year)
y_t amount of product to produce and sell in year t (tonnes)

Then the problem is to maximize the total profit achieved by selling the product less the total fixed costs:

$$\sum_t P y_t - \sum_t C_t(x_t)$$

subject to the constraints that demand must be met without exceeding the available capacity of the plant in any year:

$$y_t \qquad\qquad \leqslant D_t \quad \text{(for } t = 1, \ldots, T)$$

$$y_t - \sum_{t_1 = 1}^{t} x_{t_1} \leqslant 0 \quad \text{(for } t = 1, \ldots, T)$$

$$x_t \geqslant 0, \quad y_t \geqslant 0 \quad \text{(for } t = 1, \ldots, T)$$

However, the problem can be solved more effectively if we define nodes N_{it} representing the *state* of having i (total) units of capacity available in year t. The generalized distances a_{ijt} are then defined as:

$$a_{ijt} = \infty \qquad \text{(for } j < i)$$

since we never lose capacity. Otherwise,

$$a_{ijt} = C_t(j - 1) - P \min(D_t, j)$$

We also introduce dummy arcs of zero length from all nodes N to the finishing node, since we take no interest in what happens after T years.

This model can be refined in various ways, such as discounting future costs and revenues compared with present ones, without affecting the computational requirements at all significantly. However, the model does rely on the fact that, for all future purposes, the only thing that matters about the present state is the total capacity installed. Its age is not relevant, nor is the question of whether it was commisioned all at once or in stages. Therefore we cannot represent the condition that capacity becomes obsolete after some fixed number of years, nor can we allow variable operating costs to depend both on the level of production and on the type of plant, although fixed operating costs are associated with the activity of installing the capacity and can take any form.

12.4   THE ALLOCATION OF EFFORT

An important application of dynamic programming in a context which is not obviously dynamic is the *allocation of effort*. Suppose that we have to choose non-negative integers x_j (for $j = 1, \ldots, J$) to maximize

$$\sum_j f_j(x_j)$$

subject to the single constraint

$$\sum_j x_j \leqslant T$$

For example, we may have up to T people to allocate to J jobs, where the benefit or profit gained from x_j people working on job j is given by $f_j(x_j)$.

This problem can be solved by elementary means if the functions $f_j(x_j)$ are all convex or if they are all concave. Otherwise dynamic programming is useful because, although this is not obviously a shortest-route problem, the fact that the objective function is a sum of arbitrary functions of single activities suggests that we may be able to define a network such that these activities represent arcs.

For example, we can define nodes N_{nj} representing the allocation of n units of resource to the first j activities. Then the shortest distance from the origin to node N_{nj} represents

$$- \max \sum_{i=1}^{j} f_i(x_i)$$

subject to the constraint that

$$\sum_{i=1}^{j} x_i = n$$

It is now clear that

$$D_{nj} = \min_{x_j} \left(- f_j(x_j) + D_{n-x_j, j-1} \right)$$

or, in other words, the arc from node N_{xj-1} to N_{yj-1} has length $- f_j(y - x)$.

12.5   THE OPTIMUM DIVISION OF A LINE INTO SEGMENTS

Another general problem that arises in various contexts is to find the optimum way to divide a line into segments or subintervals. One possible application might be a railway line which must be divided into segments for maintenance of the railway points from various depots along the line. The cost of maintaining a complete segment from a single depot may then be a constant plus the weighted sum of distances of the points on the line from the depot. The weights are given as input data for each of a (finite) number of points on the line.

Therefore suppose that there are N possible *break points* where we might divide the line into segments. Suppose further that we can define or compute numbers a_{ij} representing the cost of treating the line from possible break point i to possible break point j as a complete single segment. When i is 0 this cost refers to the segment beginning at the start of the line and going to break point j, and when $j = N + 1$ this refers to the segment from the ith break point to the end of the line.

The whole problem now reduces to finding the shortest route from node 0 to node $N + 1$ through a network where the distance from node i to node j is a_{ij}. This can be computed very easily by defining D_j as the shortest distance from node 0 to node j, with P_j as the predecessor node. We then have the recurrence relationship

$$D_j = \min_{i; i < j} (a_{ij} + D_i)$$

where P_j is the corresponding value of i.

References

ABADIE, J., and CARPENTIER, J. (1969). 'Generalisation of the Wolfe reduced gradient method to the case of nonlinear constraints.' In *Optimization* (Ed. R. Fletcher), Academic Press, London and New York.

BEALE, E. M. L. (1968). *Mathematical Programming in Practice*, Pitman Publishing, London.

BEALE, E. M. L. (1975). 'The current algorithmic scope of mathematical programming systems.' *Mathematical Programming Study 4*, pp. 1–11.

BEALE, E. M. L. (1977). 'Integer programming.' In *The State of the Art in Numerical Analysis* (Ed. D. A. H. Jacobs), Academic Press, London and New York, pp. 409–48.

BEALE, E. M. L. (1985). 'The evolution of mathematical programming systems.' *J. Opl Res. Soc.* **36**, No. 5, 357–66.

BEALE, E. M. L., BEARE, G. C., and BRYAN–TATHAM, P. (1974). 'The DOAE reinforcement and redeployment study: a case study in mathematical programming.' In *Mathematical Programming in Theory and Practice* (Eds P. L. Hammer and G. Zoutendijk), North-Holland, Amsterdam, pp. 417–42.

BEALE, E. M. L., and TOMLIN, J. A. (1970). 'Special facilities in a general mathematical programming system for non-convex problems using ordered sets of variables.' In *Proceedings of the Fifth International Conference on Operational Research* (Ed. J. Lawrence), Tavistock Publications, London, pp. 447–54.

BELLMAN, R. (1957). *Dynamic Programming*, Princeton University Press, Princeton, New Jersey.

BRENT, R. P. (1971). 'An algorithm with guaranteed convergence for finding a zero of a function.' *Computer Journal* **14**, 422–5.

BROYDEN, C. G. (1970). 'The convergence of a class of double rank minimization algorithms,' parts I and II. *J. Inst. Maths Applns* **6**, 76–90 and 222–31.

DANTZIG, G. B. (1951). 'Maximization of a linear function of variables subject to linear inequalities.' In *Activity Analysis of Production and Allocation* (Ed. T. C. Koopmans), John Wiley, New York.

DANTZIG, G. B. (1955). 'Upper bounds, secondary constraints and block triangularity.' *Econometrica* **23**, 174–83.

DANTZIG, G. B. (1963). *Linear Programming and Extensions*, Princeton University Press, Princeton, New Jersey.

DANTZIG, G. B., and ORCHARD HAYS, W. (1954). 'The product form of inverse in the simplex method.' *Math. Comp.* **8**, 64–7.

DANTZIG, G. B., ORDEN, A., and WOLFE, P. (1955). 'The generalized simplex method for minimizing a linear form under linear inequality restraints.' *Pacific Journal of Mathematics* **5**, 183–95.

DANTZIG, G. B., and VAN SLYKE, R. (1967). 'Generalized upper bounding techniques.' *J. Comp. Systems Science* **1**, 213–26.

DAVIES, M. (1967). 'Linear approximation using the criterion of least total deviations.' *J. Roy. Statist. Soc. (B)* **29**, 101–9.

DAVIDON, W. C. (1959). 'Variable metric method for minimization.' *AEC Res. and DEV. Report ANL-5990* (revised).

DIXON, L. C. W., and SZEGÖ, G. P. (1978). *Towards Global Optimization 2*, North-Holland, Amsterdam.

FLETCHER, R. (1970). 'A new approach to variable metric algorithms.' *Computer Journal* **13**, 317–22.

FLETCHER, R., and POWELL, M. J. D. (1963). 'A rapidly convergent descent method for minimization.' *Computer Journal* **6**, 163–8.

FLETCHER, R., and REEVES, C. M. (1964). 'Function minimization by conjugate gradients.' *Computer Journal* **7**, 149–54.

FOURER, R. (1983). 'Modeling languages versus matrix generators for linear programming.' *ACM Trans. Maths Software* **9**, 143–83.

GOLDFARB, D. (1970). 'A family of variable-metric methods derived by variational means.' *Math. Comp.* **24**, 23–6.

GOLDFELD, S. M., QUANDT, R. E., and TROTTER, H. F. (1966). 'Maximization by quadratic hill-climbing.' *Econometrica* **34**, 541–51.

GOMORY, R. E. (1958). 'Outline of an algorithm for integer solutions to linear programs.' *Bulletin of the American Mathematical Society* **64**, 275–8.

HARIS, P. M. J. (1973). 'Pivot selection methods of the Devex LP code.' *Mathematical Programming* **5**, 1–28.

HELLERMEN, E., and RARICK, D. (1971). 'Reinversion with the preassigned pivot procedure.' *Mathematical Programming* **1**, 195–215.

HESTENES, M. R., and STIEFEL, E. (1952). 'Methods of conjugate gradients for solving linear systems.' *J. Res. NBS* **49**, 409–36.

KUHN, H. W., and TUCKER, A. W. (1951). 'Non-linear programming.' *Proceedings of Second Berkeley Symposium on Mathematical Statistics and Probability* (Ed. J. Neyman), University of California Press, pp. 481–92.

LEMKE, C. E. (1954). 'The dual method of solving the linear programming problem.' *Naval Research Logistics Quarterly* **1**, 36–47.

LEVENBERG, K. (1944). 'A method for the solution of certain non-linear problems in least squares.' *Qu. App. Maths* **2**, 164.

LITTLE, J. D. E., Murty, K. C., SWEENEY, D. W., and KAREL, C. (1963). 'An algorithm for the travelling salesman problem.' *Operations Research* **11**, 972–89.

MARKOWITZ, H. M. (1957). 'The elimination form of the inverse and its application to linear programming.' *Management Science* **3**, 255–69.

MARKOWITZ, H. M., and MANNE, A. S. (1957). 'On the solution of discrete programming problems.' *Econometrica* **25**, 84–110.

MILLER, C. E. (1963). 'The simplex method for local separable programming.' In *Recent Advances in Mathematical Programming* (Eds R. L. Graves and P. Wolfe), McGraw-Hill, New York.

MURTAGH, B. A., and SAUNDERS, M. A. (1978). 'Large-scale linearly constrained optimization.' *Mathematical Programming* **14**, 41–72.

ORCHARD HAYS, W. (1978a). 'History of mathematical programming systems.' In *Design and Implementation of Optimization Software* (Ed. H. J. Greenberg), Sijthoff & Noordhoff, The Netherlands.

ORCHARD HAYS, W. (1978b). 'Scope of mathematical programming software.' In *Design and Implementation of Optimization Software* (Ed. H. J. Greenberg), Sijthoff & Noordhoff, The Netherlands.

ORCHARD HAYS, W. (1978c). 'Anatomy of a mathematical programming software.' In *Design and Implementation of Optimization Software* (Ed. H. J. Greenberg), Sijthoff & Noordhoff, The Netherlands.

PETERS, G., and WILKINSON, J. H. (1969). 'Eigenvalues of $Ax = \lambda Bx$ with band symmetric A and B.' *Computer Journal* **12**, 398–404.

POWELL, M. J. D. (1973). 'On search directions for minimization algorithms.' *Mathematical Programming* **4**, 193–201.

POWELL, M. J. D. (1977). 'Restart procedures for the conjugate gradient method.' *Mathematical Programming* **12**, 241–54.

POWELL, M. J. D. (Ed.) (1982). *Nonlinear Optimization 1981*, Academic Press, London and New York.

SCHRAGE, L. (1975). 'Implicit representation of variable upper bounds in linear programming.' *Mathematical Programming Study* **4**, 118–32.

SHANNO, D. F. (1970). 'Conditioning of quasi-Newton methods for function minimization.' *Math. Comp.* **24**, 647–56.

VAN WIJNGAARDEN, A., ZONNEVELD, J. A., and DIJKSTRA, E. W. (1963). *Programs AP200 and AP230 De serie AP200* (Ed. T. J. Dekker), The Mathematical Centre, Amsterdam.

WILLIAMS, H. P. (1978). *Model Building in Mathematical Programming*, John Wiley, Chichester.

WOLFE, P. (1963). 'Methods of nonlinear programming.' In *Recent Advances in Mathematical Programming* (Eds R. L. Graves and P. Wolfe), McGraw-Hill, New York.

WOLFE, P. (1965). 'The composite simplex algorithm.' *SIAM Review* **7**, 42–54.

Index